"Concise in expression, with handsome cartography and a text that sticks to the facts, it elucidates the key issues surrounding global climate change." *Kirkus Review*

"Seeing that climate change is probably the biggest single problem facing our world today, it's natural that you might want to find out more. You could wade through dense, dry academic detail from the IPCC. Or you could root out the newly published *Atlas of Climate Change,* which condenses key findings from the scientists and is packed with facts, graphs, and maps explaining how climate change is affecting the planet." *The Guardian*

'This is a remarkable piece of work and extremely readable. What is heartening is not only the wealth of information it conveys but also the powerful illustrations and graphics." **Director-General, The Energy and Resources Institute (TERI)**

"This pioneering atlas will become an essential point of reference for anyone looking for a quick and accurate overview of this multidisciplinary subject."
Foreign and Commonwealth Office, UK

"An excellent book and wonderful tool to help clarify the mountains of data into something comprehensible. If you want to get up to speed on the subject, this is probably the best place to start. Highly recommended." *Transition Culture*

"An excellent vehicle for beginning to understand the climate change issue."
The Forestry Chronicle, **Canada**

"Packs a lot of very high quality information into a small space with an excellent layout. A hit with students and essential for every library shelf – one of the best buys of the year." *Ecological and Environmental Education*

"If you want to understand the evidence, causes, and consequences of climate change, as well as what we can do to deal with it, get your copy today." *CarbonSense*

"A succinct, yet exacting new resource that will be vital for many years to come."
Oxford University Centre for the Environment

"This vast mine of information takes a wider than usual view of the implications of climate change. It is a valuable resource to help understand the core issues and puts into perspective the desperate unfairness of the global picture." *Morning Star*

"All schools and colleges should complement their environment studies with this atlas." *Education Journal*

In the same series:

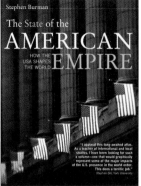

THE STATE OF THE AMERICAN EMPIRE
by Stephen Burman

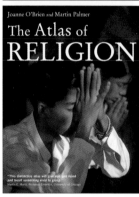

THE ATLAS OF RELIGION
by Joanne O'Brien
and Martin Palmer

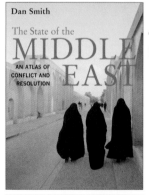

THE STATE OF THE MIDDLE EAST
by Dan Smith

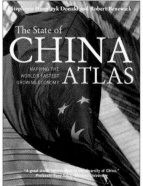

THE STATE OF CHINA ATLAS
by Stephanie Hemelryk Donald
and Robert Benewick

THE ATLAS OF
CLIMATE CHANGE

MAPPING THE WORLD'S GREATEST CHALLENGE

Revised and Updated

Kirstin Dow and Thomas E. Downing

UNIVERSITY OF CALIFORNIA PRESS

Berkeley   Los Angeles

University of California Press, one of the most distinguished university presses in the United States, enriches lives around the world by advancing scholarship in the humanities, social sciences, and natural sciences. Its activities are supported by the UC Press Foundation and by philanthropic contributions from individuals and institutions. For more information, visit www.ucpress.edu.

University of California Press
Berkeley and Los Angeles, California

The moral right of the author has been asserted.

ISBN 978-0-520-25558-6

The Library of Congress has cataloged an earlier edition of this book as follows:

Library of Congress Cataloging-in-Publication Data

Dow, Kirstin, 1963-
The atlas of climate change : mapping the world's greatest
challenge / Kirstin Dow and Thomas E. Downing.
p.   cm.
Includes bibliographical references and index.
ISBN-13 978-0-520-25023-9 (pbk. : alk. paper)
ISBN-10 0-520-25023-0 (pbk. : alk. paper)
1. Climatic changes.   1. Climatic changes--Maps.
3. Climatic changes--Charts, diagrams, etc.
I. Downing, Thomas E.   II. Title.

QC981.8.C5D69   2006
551.6--dc22   2006050098

Produced for the University of California Press by
Myriad Editions
59 Lansdowne Place
Brighton, BN3 1FL, UK
www.MyriadEditions.com

Edited and co-ordinated by Jannet King and Candida Lacey
Maps and graphics created by Isabelle Lewis
Design and additional graphics by Corinne Pearlman

Printed on paper produced from sustainable sources.
Printed and bound in Hong Kong through Phoenix Offset Limited
under the supervision of Bob Cassels, The Hanway Press, London.

14   13   12   11   10   09   08   07
10   9   8   7   6   5   4   3   2   1

Contents

Foreword by Bo Kjellén   8
Authors   8
Introduction   9
Definition of Key Terms   14

PART 1: SIGNS OF CHANGE   19

Warning Signs
*New records and observations around the world are
consistent with scientists' expectations of climate change.* 20
Polar Changes
*Warming in the polar regions is driving large-scale melting
of ice that will have both local and global consequences.*   22
Glacial Retreat
*Most of the world's glaciers are retreating at unprecedented
rates – a clear sign of warming.*   24
Everyday Extremes
*Weather-related disasters are becoming increasingly
common around the world.*   26

PART 2: FORCING CHANGE   29

The Greenhouse Effect
*The increasing concentration of greenhouse gases is
trapping more heat.*   30
The Climate System
*The entire climate system is adjusting to an increase in the
heat trapped in the Earth's atmosphere.*   32
Interpreting Past Climates
*Concentrations of carbon dioxide and methane are higher
than they have ever been in the last 650,000 years. The
Earth is warmer than in the past 1,000 years.*   34
Forecasting Future Climates
Global temperatures are predicted to continue rising.   36

PART 3: DRIVING CLIMATE CHANGE   39

Emissions Past and Present
*Most greenhouse gases have been, and are, emitted to meet
the needs of modern industrial societies.*   40
Fossil Fuels
*The emission of greenhouse gases from the burning of fossil
fuels is the major cause of climate change.*   42
Methane and Other Gases
A range of greenhouse gases contribute to climate change.   44
Transportation
*International trade, travel and a growing dependence on
motor vehicles make transportation one of the main sources
of greenhouse gas emissions.*   46

Disrupting the Carbon Balance
*Carbon is essential in the natural environment, but changes
in land use may release stored carbon and contribute to
climate change.*   48

Agriculture
*Greenhouse gases are emitted in the production of food.
While some agriculture meets basic needs, some simply
provides wealthy consumers with the luxury of choice.*   50

PART 4: EXPECTED CONSEQUENCES   53

Disrupted Ecosystems
*Many species and ecosystems, already at risk from human
development, may not be able to adapt to new climatic
conditions and stresses.*   54

Threatened Water Supplies
*Water scarcity is already a growing concern. In some places
climate change will make it even more critical.*   56

Food Security
*Climate change threatens food security, although crop yields
in temperate regions may improve.*   58

Threats to Health
*Climate change threatens human health. The poorest
regions are likely to be the hardest hit.*   60

Rising Sea Levels
*Thermal expansion of oceans and melting ice will lead to a
substantial rise in sea level, threatening many coastal communities.*   62

Cities at Risk
*Coastal erosion, salt-water intrusion into freshwater supplies,
and coastal storms all threaten coastal areas – often regions
of high population growth and intensive economic development.*   64

Cultural Losses
*Damage to indigenous cultures, historical monuments and
archaeological sites adds to the incalculable economic
losses of climate change.*   66

PART 5: RESPONDING TO CHANGE   69

International Action
*Most countries have acknowledged the problem of climate
change by signing the Convention on Climate Change.*   70

Meeting Kyoto Targets
*Many countries are making progress towards their Kyoto
commitments, but even the agreed targets fall far short of
stabilizing greenhouse gas emissions at levels considered to be safe.*   72

Carbon Trading
*Trading in carbon credits is one way to share the burden of
reducing emissions globally.*   74

Financing Responses
Current funding is inadequate to help countries respond to climate change. 76

Local Commitment
In many places, local and regional authorities are developing more aggressive emission reduction policies than federal governments. 78

Carbon Dioxide and Economic Growth
Economic growth can be achieved with lower greenhouse gas emissions. 80

Renewable Energy
Renewable energy sources could be the technological key to economically and socially sustainable societies. 82

Adapting to Change
The capacity to adapt to climatic hazards and stresses depends on a country's wealth, resources and governance. 84

PART 6: COMMITTING TO SOLUTIONS 87

Personal Action
People all over the world are taking measures to reduce the greenhouse gases emitted as a result of the way they live. 88

Public Action
The policies, practices, and investments of governments, businesses, and civic organizations will have the greatest impact on our future. 90

PART 7: CLIMATE CHANGE DATA 93
Data table 94

Sources 102
Index 110

Foreword

We live in a confusing and rapidly changing world. Trends are appearing and disappearing in arts, fashion, lifestyles, and ideas, at breathtaking speed. New information and communication technologies open new cultural and commercial opportunities. At the same time economic and social inequalities are growing.

We also face challenges of another kind, never experienced before. A world population of six billion people with unprecedented technological capacity now has an impact on the global natural system. This has led Paul Crutzen and Eugene Stoermer to argue that we live in a new geological period, the Anthropocene era. Humans have become a force of nature, triggering changes in immense ecosystems – changes which could rebound, threatening our own livelihoods, and the lives of future generations.

Climate change is the prime example of this process. Global warming is caused by greenhouse gases being released into the atmosphere, where no national borders exist. The most important greenhouse gas, carbon dioxide, is emitted in energy production and transportation – which have jointly formed the basis of industrial development over the last 200 years. Industrial production, based on cheap energy, has enabled recent generations to enjoy luxuries of life never before experienced. For most of us in the developed world, the private car has become a symbol of freedom and new opportunity.

It is not surprising that the arguments and warnings of climate scientists create controversy. There is much at stake. Do they exaggerate the dangers? What about natural variations in climate? Does not action against climate change become too costly? Democratically elected political leaders have concluded that action is necessary, that the threat of climate change needs to be taken seriously, but the position taken by the USA on the Kyoto Protocol shows that there are still serious disagreements about the structure of the international agreements needed to tackle the threat. Decisions taken by the Conference of Parties in Montreal late in 2005 have opened the way for a serious discussion of the climate regime after 2012, but the process is still uncertain and precarious.

The question of the future of oil resources has introduced a new dimension to these discussions. When will the moment of peak oil appear? Are there any unknown resources? How can the world community handle the foreign policy and security implications of this state of affairs?

This is indeed a time for careful reflection at all levels. Climate change has become a highly political issue, involving very large economic interests. The long-term character of the problems is daunting; investment decisions will have consequences for infrastructure and energy systems that will stretch over many decades. Social tensions may appear in the wake of action to counter climate change.

It is in this perspective that this atlas provides essential background reading. It provides the facts, enabling readers to form an independent view of the problems, permitting them to judge the action of governments and international organizations in a realistic and rational way. Climate change action must be based on the support of a well-informed public, ready to accept bold policies and treaties.

Bo Kjellén
Former Swedish Ambassador and Chief Negotiator
for the UN Framework Convention on Climate Change

Authors

Dr. Kirstin Dow is Associate Professor, University of South Carolina Department of Geography and Senior Research Fellow, Stockholm Environment Institute. She conducts research on topics related to climate, vulnerability, environmental change, and society. She is a National Councillor of the Association of American Geographers, a contributor to the Millennium Ecosystem Assessment, and a Principal Investigator in NOAA's Regional Integrated Science and Assessment (RISA) network, addressing climate research and services.

Dr. Thomas E. Downing is Director, Stockholm Environment Institute, Oxford Office, and Visiting Fellow, Queen Elizabeth House, University of Oxford. He is Munich Re Foundation Chair (designate) in social vulnerability with the United Nations University Institute for Environment and Human Security. He carries out research on topics related to climate and society. He contributes to the Intergovernmental Panel on Climate Change, is the past chair of the International Geographical Union's Task Force on Vulnerability, and has been advisor to the UK Climate Impacts Programme and House of Commons International Development Committee.

Introduction

The broad-reaching effects of climate change are becoming more apparent to scientists, and to people in their daily lives. Residents in parts of Alaska have already experienced warming of up to 4°C, with buildings subsiding as the ground beneath them thaws for the first time in memory. Gardeners and bird watchers are observing longer growing seasons, altered migratory patterns, and earlier blooming. More extreme weather patterns, bringing heatwaves, droughts and floods, seem increasingly common. Old photographs retain the images of glaciers, seemingly invulnerable, that are now gone. Around the world, the predictions of climate change – that we would much rather dismiss – are now all too obvious. It is common to hear people joke that unusual weather is due to climate change; such remarks are no longer uttered entirely in jest.

The rate of climate change is unprecedented, and the change is almost certain to reach levels that severely challenge environmental management and international governance capabilities. It will take new institutional strategies and forms of cooperation, and a willingness to deal with longer time frames in decision making. As scientists who, for years, have been studying the consequences of climate and other environmental changes on people and livelihoods, we believe climate change and its potential impacts are extremely serious issues, and that some expected impacts are very dangerous. Other consequences may not be as dangerous, but concerted efforts will be required to adapt to the changes. We also believe that a better understanding of the problem and its complexities will help us to avoid some of the threats, direct our efforts wisely, and find opportunities for meeting the challenges. To date, climate policy has been largely in the hands of scientists, environmental activists and politicians. But it is time to bring more minds and perspectives to the task. The required changes will involve us all deeply.

After listening to shrill warnings of catastrophe, dismissive statements from skeptics, decades of calls for more research, incomplete stories of the complexities of negotiations, and confusion over the level of consensus in scientific understanding, we want to provide a resource to help people make that initial step into understanding the core issues of climate change. In this atlas, each map spread introduces central issues in current debates about the physical processes, the likely impacts, and developing adequate responses.

A vast knowledge, touching every aspect of science and life, is emerging from decades of engagement and debate on climate change. This understanding does not eliminate uncertainty altogether; but amid the complexity, uniqueness, and variability, there are topics of consensus, supported by large bodies of scientific literature. The publications, early in 2007, of the Fourth Assessment Report by the Intergovernmental Panel on Climate Change, illustrate the improved understanding of climate science, regional impacts and response strategies, based on the tremendous growth of evidence from around the world. This atlas was originally compiled while the IPCC's Working Groups were still making their assessment, and although we had largely anticipated their findings, we took the opportunity in 2007 to make some amendments in the light of the summary reports of Working Groups I, II and III. The following discussion synthesizes our analysis, supporting our view that we know enough to act, and will face ever more serious consequences for delays in doing so.

Monitoring change

Evidence shows that human-induced climate change is taking place, today. The fundamental understanding of the physics of atmospheric gases in determining Earth's energy balance and affecting global temperatures has been established for over 100 years. There is no doubt that human activities have changed the chemical composition of the atmosphere. Models based on these facts are generating output consistent with observations. Carbon dioxide levels are higher today than at any time in the last 650,000 years. The melting of the polar ice caps, thawing of northern permafrost, and retreat of glaciers match expectations for greater impacts in the polar regions. More frequent occurrence of drought and heatwave disasters is consistent with the predictions of changes in the average range of variability. Across the planet, birds, butterflies and other species are changing their ranges in response to climate signals. In addition, while these models capture the climate systems well enough to replicate Earth's climate history, they cannot account for current trends of global warming without incorporating the added burden of greenhouse gas emissions caused by human actions. When the models work forward, they project more warming, rising sea levels, shifting regional patterns of temperature and precipitation. And, due to the scale of processes undergoing change, it is apparent that past and future carbon dioxide emissions will continue to influence warming and sea-level rise for centuries to come.

It is the details of these shifts within the climate system which are at the center of current scientific debate. Unfortunately, we can no longer hold on to hopes that the climate change skeptics were right. That period of skepticism of the fundamentals has ended and been replaced by debates over specific impacts: the likely rate of change, the amount of sea level rise, the probability of the collapse of the West Antarctic ice sheet, and the impacts on hurricane frequency and intensity. There are also important debates on the economically and socially optimal ways of mitigating and adapting to change, as well as the best strategies for intergovernmental cooperation.

Understanding the implications

In order to anticipate the future, we have had to examine the present in a new light. The effort put towards understanding the implications of climate change has had the somewhat counterintuitive result of making more apparent the importance of climate in our lives today. From global health to water supplies, food security, and coastline development, all societies are more dependent on climate than is often appreciated. How and where we build, plant, harvest, commute, work, and play all reflect expectations of climate conditions.

In some places, this examination makes it abundantly clear that people, economies, and ecosystems are at serious risk from current patterns of climate variability. Across large parts of Africa and Asia, the timing and the abundance of rains determines whether crops will support households, or whether hungry people will need to search for alternative sources of food and income. Sometimes, these dry periods last for years. The poor – among them young children, pregnant women, the elderly, and the sick – are most severely affected, and many thousands die. Where other disasters, such as HIV/AIDS, already afflict households, diminished resources create greater hardships. As we learned from the 2003 heatwave in Europe, even the most advanced industrialized countries face serious risks.

Around the world and at a local level, climate conditions are also critical to the distribution and wellbeing of ecosystems. Many ecosystems are currently under tremendous pressure from land-use change, pollution, and over-harvesting. Present demands are altering the character of entire systems, from the number and variety of species present to disease sensitivity and the ability to provide ecosystem services, such as water filtration and flood control. The only remaining examples of some ecosystems exist in carefully delimited and protected areas, where the authorities struggle to manage the pervasive impacts of wide-ranging processes, including atmospheric deposition and watershed changes. Climate change is emerging as another unavoidable stress, as species begin to shift their ranges, and altering patterns of temperature and precipitation influence the dynamics of local systems. Scientists have analyzed over 29,000 series of observations, drawn from 75 studies of physical and biological systems around the world, and found that around 90 percent of changes in these systems are consistent with the direction of change expected as a response to warming. But the ability of entire ecosystems to shift is uncertain, both because the rate of change is faster than typical ecological timescales, and because there is limited space available to move. The area immediately surrounding protected areas and coastal areas is often already in use and, in the far north and on mountain tops, there are no more options.

On a larger scale, the distribution of current climate benefits, risks, and debates over responsibility are shaped by the global economic system. On a per capita basis, both historically and currently, wealthy industrialized nations benefit from activities that release a disproportionate level of greenhouse gas emissions. Globalization, with its impacts on industrial location, economic development, and extensive transport demands, is influencing future patterns of emissions. Negotiating reductions is complicated by the combination of historical versus anticipated contributions associated with economic development, and export relations that result in one country producing goods and emissions for markets in another.

In addition, not all serious climate impacts will be the direct result of local changes, and even now, the interconnectedness of the global economy is a major mechanism for transferring those impacts. For instance, favorable growing conditions worldwide and a bumper crop might translate into lower prices for all producers, while a crop loss for one area may create an economic windfall for another. Similarly, countries and companies are already working to lead and to profit from investments in new technologies.

Suffering the consequences

Looking ahead, while there will be benefits for some, there will be severe consequences for others. Those presently most vulnerable – poor people in marginal areas – are likely to suffer first. Currently, over 2 billion people struggle to live on less than a dollar a day, often depending heavily on agriculture, fishing, and animal husbandry to maintain their livelihoods and hopes for better lives. With 2°C warming, millions more people, many of them among the poorest, are likely to be exposed to annual coastal flooding. Changing precipitation patterns, either wetter or drier, along with altered temperature, will affect crop productivity, availability of food and water for livestock, as well as feeding locations of fish. Climate changes may also facilitate the movement of human, plant, and animal diseases into areas where they were previously little known and where doctors, veterinarians, agriculture extension specialists, and money for treatments are all in short supply.

The urban poor, who are increasing in number, are not exempt. Many living in cities have their primary homes in rural areas and work in cities in order to support their families at home. Were climate change to cause greater resource scarcity, they would feel the costs of rising food, energy, and water costs most acutely. Some of that loss would also be shared with family in the countryside. Heat waves may make manual work unbearable. Because urban poor often live in marginal places in cities – slums along river levees or perched on steep slopes – the natural hazards of flooding, intense rainfall, and mudslides are likely to hit them hardest.

These are not the people driving climate change. Of those living on less than a dollar a day, few have electricity, cars, refrigerators, or water heaters. But, because their lives are tied to climate conditions and they have few resources to buffer against bad or progressively difficult conditions, they are likely to bear the highest human costs. Climate change is as much a humanitarian and human development concern as it is an environmental one.

As implementation of strategies to reduce greenhouse gas emissions and adapt to climate change progresses, the social and economic impacts of these programs also demand careful attention. As we write, the concerns of the current US government over the potential economic impacts of pursuing the Kyoto Protocol are widely known, as is their refusal to ratify the protocol. On a local and individual scale, the impacts of policies, such as investing in new and potentially more expensive energy systems, taxing fossil fuels, or diverting some government funds to mitigation, hazard preparedness and recovery, may weigh disproportionately on the poor around the world. In the wake of Hurricane Katrina, New Orleans cut funding for education, police, social, legal and other services. While Katrina is not attributed to climate change, it serves as another warning that even a wealthy, well prepared nation may anticipate billions of dollars of damages and that the poor will bear much of the burden.

Developing adequate responses

Responding to climate change and the threatened consequences is a very serious and sobering business, but it is not hopeless, nor are all foreseeable outcomes inevitable. Both mitigation (reducing emissions) and adaptation (adjusting to changing conditions) are essential, and there are feasible, effective, solutions. In addition, many of these actions save money and make good sense for other reasons as they also address the broader issues of sustainability through reducing dependency on increasingly scarce petroleum supplies, cutting air pollution, and preserving the ecosystems that help maintain the ecological processes we rely on. Although it is tempting to focus on technological solutions, it is essential also to address the issues of consumption and lifestyle that drive technologies.

Many countries are likely to meet their emission reduction commitments under the Kyoto Protocol, but those commitments are only the beginning of 60 to 80 percent reductions required. And, there is substantial reason to move quickly. The climate system changes slowly. The benefits of actions to reduce emissions taken today will not be realized for some decades.

Technological change alone cannot achieve the

necessary levels of emissions reductions; many currently available technologies can, however, make a substantial difference. Renewable energy technologies are becoming more diverse, affordable, and widely adopted. Photovoltaics, solar thermal heating of water, geothermal, wind, tide, small-scale hydropower, and biofuels are rapidly expanding and establishing their viability as major energy sources. These technologies offer great potential for greenhouse gas reduction as they currently account for less than 5 percent of global energy production. Automotive manufacturers, like Toyota, are demonstrating that cars can achieve much greater efficiencies. Greater efficiencies are possible in other manufacturing and power production systems as well. Interest in nuclear power has also risen, but its future role in addressing climate change issues is less certain. Concerns over safety and long term storage of radioactive wastes remain and it is not clear that its potential value as a response to climate change offers sufficiently strong justification to overcome other economic barriers.

In the building trade, new construction following the US Green Building Council's Leadership in Energy and Environmental Design (LEED) certification program is demonstrating that it is possible to design buildings that are 30 percent or more energy efficient without increasing the initial construction costs. But, realizing the full potential in all of these areas will require research and development investment to fully investigate and design technologies to take advantage of renewable, non-greenhouse gas producing energy sources. It will also involve adjustments to infrastructure and institutional practices, such as building design and training of architects and engineers. Yet, without changes in consumer demand, such as reducing the desire for ever larger homes and more powerful cars, these advances and opportunities could be fruitless.

The necessary institutional innovations depend on financing for new technologies and for replacing inefficient power plants and industrial processes. The establishment of carbon trading markets is a promising example. These markets are designed to share the burden of emissions reduction by providing countries and companies with greater flexibility in how they meet reduction targets, and enabling them to benefit from making changes early. Developing certification schemes to document reductions, bringing buyers and

sellers of carbon credits together, and other challenges are now being addressed. Markets, such as the EU trading scheme and the Regional Greenhouse Gas Initiative in the USA are getting underway. The level of trade is growing rapidly, but still represents a very small percentage of world carbon emissions.

Signatories to the UN Framework Convention on Climate Change, most of the countries in the world, have committed to preventing "dangerous anthropogenic [human-induced] climate change". It is clear that reducing emissions is not a sufficient response in the face of the potentially dangerous consequences of climate change. Adaptation is essential.

All levels of effort, from individuals to cities to nations, corporations, and the international community are needed to meet this challenge. From individuals taking the 1-tonne emissions reduction challenge to business leadership councils offering toolkits for reductions, to strong national programs and active international negotiations, this engagement is taking place. Many more efforts are underway to support broadening participation. Where federal commitment to international cooperation is viewed as lagging, as in the USA and Australia, hundreds of cities have committed to move more quickly to reduce emissions reductions by using renewables, choosing less carbon intensive technologies, implementing more efficient urban design, and increasing energy efficiency. Businesses are setting their own targets, and exceeding them, even saving money at the same time.

Balancing commitment with responsibility is difficult in designing "fair" cooperative efforts. The industrialized nations are responsible for the majority of historical emissions as well a disproportionate amount of the current emissions. In some nations, reductions might come from limiting luxuries, like numbers of very large cars, while in others the majority of emissions are associated with agriculture, and commitments to large reductions would cut into food production. Many countries lack the resources to provide basic services, much less make substantial investments in mitigation and adaptation. Some international funding is available, but it is less than 2 percent of the Official Development Assistance budget, already under-funded to meet established commitments.

Action on climate change is a part of diverse global movements. It is grounded in environmental science and action, and in concerns for sustainability. It is often linked to other environmental issues, such as ozone depletion, or economic trends, such as the peak in oil production (which may have already occurred). Our experience of climate change is intimately exposed in the record of natural disasters. And disaster reduction has a humanitarian urgency as we seemingly lurch from crisis to crisis.

These roots in environment, economy, and disasters are related to the concern for ending poverty, from the halfway there targets of the Millennium Development Goals to the popular voices of Live 8. Governance and economic growth are guiding lights, but so are equity and fair trade.

We believe solutions to the challenges and opportunities of climate change must reflect each of these goals. We encourage you to get involved, in your own lives, in your own, even virtual, communities, and in our collective endeavors. In the near future, we hope to have more stories of successful mitigation and adaptation to report. Please send yours to: ClimateAtlas@gmail.com

Acknowledgements

Throughout this project, we were fortunate to have the support and expertise of family, friends, and colleagues. It meant a great deal to us personally. It was also indispensable in allowing us to create this resource for understanding the challenges of climate change.

No project like this goes to print without long nights, working weekends, late dinners, occasional missed appointments, and additional travel as we chose the key messages, hunted for reliable sources, honed the text and debated minute details. We thank our families and friends for their patience with all of this and more.

Friends and colleagues contributed insight and data from their own research, developed custom graphics, consulted on highly specialized topics, investigated the deepest details of data sets, provided office space, reviewed drafts, and earned our deep gratitude as well as many IOUs. Specific contributors include: Roger Barry, Ruth Butterfield, Greg Carbone, Carl Dahlman, Jason Downing, Jonathan Downing, Stephan Harrison, Bushra Hassan, Saleemul Huq, Geoff Jenkins, Nayna Jhaveri, Richard J T Klein, Neil Leary, Adam Lofts, Molly McAllister, Kimberly Meitzen, Cary Mock, Susi Moser, Adrian Rodger, Fiona Smith, David Stainforth, Athanasios Vafeidis and Richard Williams.

The team at Myriad Editions brought the data to life with their graphical imaginations, and worked to assure that the technical details and text were as clear and current as possible for this rapidly evolving topic. Candida Lacey sold us the concept, and marketed the product. Jannet King persevered with us through the thick and thin of deadlines, the stickiest of data issues, some atmospheric chemistry, scientific debates, and other challenges. Isabelle Lewis, with assistance from Corinne Pearlman, created essential graphics and unique maps. The knowledgeable staff of Earthscan provided careful readings that led to greater clarity. Phil O'Keefe and Chris West offered thoughtful reviews of the manuscript.

It all started with Joni Seager, and Roger Kasperson brought us together in the Stockholm Environment Institute.

As this atlas is a synthesis of decades of scientific research from around the world, we also benefited greatly from the commitment of many scientists we have not met. We developed a renewed and deeper appreciation for the challenges and effort involved in developing rigorous international databases, scientific publication on the many facets of climate change, and communicating to a more general audience. We have struggled to keep up with their productivity and make the atlas as current as possible. The print deadline prevented us adding new material beyond Spring 2006, although we are pleased to have had an opportunity, with this reprint of the first edition, to make some minor updates following publication of the IPCC reports in 2007. With that caveat, all remaining shortcomings and errors are ours.

Thanks to you all. We hope this atlas encourages you in your individual efforts and our collective purpose.

Kirstin Dow and Tom Downing
May 2007

Definition of Key Terms

Definitions for chemical names, units, technical terms, and regional groupings recognized in international treaties are provided below, along with explanatory notes on impacts, vulnerability and adaptation, theory, predictions, forecasts and scenarios, and the IPCC suite of scenarios. Sources for the definitions are provided at the end of the book.

Chemical names

CCl_4 Carbon tetrachloride.

CFC Chlorofluorocarbon – covered under the 1987 *Montreal Protocol on Substances that Deplete the Ozone Layer* and used for refrigeration, air conditioning, packaging, insulation, solvents, or aerosol propellants. Since they are not destroyed in the lower **atmosphere**, CFCs drift into the upper atmosphere where, given suitable conditions, they break down **ozone**. These gases are being replaced by other compounds, including **HCFCs** and **HFCs**, which are greenhouse gases covered under the **Kyoto Protocol**.

CH_4 Methane.

CO_2 Carbon dioxide.

Halocarbons Compounds containing carbon and one or more of the three halogens: fluorine, chlorine, and bromine, including the greenhouse gases CFCs, CCl_4, and HFCs.

HCFC Hydrochlorofluorocarbon.

HFC Hydrofluorocarbon.

N_2O Nitrous oxide.

O_3 Ozone in the lower atmosphere (troposphere) that acts as a **greenhouse gas**. It is created both naturally and by photochemical reactions involving gases resulting from human activities ("smog"). In the stratosphere, ozone is created by the interaction between solar ultraviolet radiation and molecular oxygen (O_2). In the upper atmosphere (stratosphere) ozone plays a decisive role in the stratospheric radiative balance. Its concentration is highest in the ozone layer. Depletion of stratospheric ozone results in increased ultraviolet radiation.

PFCs Perfluorocarbons.

SF_6 Sulfur hexafluoride.

Units

GW Gigawatt

GWth Gigawatt thermal

Tonnes Metric tons, equivalent to 1,000 kg or 2,204.62 pounds, A gigatonne is one billion (10^9) tonnes.

Joule A Standard International (SI) unit of energy; 1,055 Joules = 1 BTU; 1 Joule = 0.2389 calories. A petajoule is 1 quadrillion (10 to the power of 15) joules.

Technical terms

Anaerobic A life or process that occurs in, or is not destroyed by, the absence of oxygen.

Anthropogenic Resulting from or produced by human beings.

Calving The breaking away of a mass of ice from a floating glacier, ice front, or iceberg.

Carbon dioxide equivalent (CO_2e) A measure used to compare the emissions from various greenhouse gases, based on their global warming potential (GWP). The carbon dioxide equivalent for a gas is derived by multiplying the tonnes of the gas by the associated GWP.

Carbon equivalent A measure used to compare the emissions of different greenhouse gases, based on their global warming potential (GWP), arrived at by multiplying carbon dioxide equivalents by 12/44.

Carbon sequestration The removal and storage of carbon from the atmosphere in **carbon sinks**.

Carbon sink Reservoirs for carbon, such as forests and oceans, processes, activity or mechanisms that store more carbon than they release.

CDM Defined in Article 12 of the Kyoto Protocol, the Clean Development Mechanism is intended to meet two objectives: (1) to assist Parties not included in Annex I in achieving sustainable development and in contributing to the ultimate objective of the convention; and (2) to assist Parties included in Annex I in achieving compliance with their quantified emission limitation and reduction commitments.

Climate In a narrow sense this is usually defined as the "average weather" or the statistical description of the mean and variability of relevant quantities over a period of time ranging from months to thousands or millions of years. The classical period is 30 years, as defined by the World Meteorological Organization (WMO). The relevant quantities are most often surface variables such as temperature and precipitation. Climate in a wider sense is the state of the climate system.

Climate change A statistically significant variation in either the mean state of the **climate** or in its variability, persisting for an extended period (typically decades or longer). Climate change may be due to natural internal processes or external **radiative forcing**, or to persistent **anthropogenic** changes in the composition of the atmosphere or in land use. The **UNFCCC**, in its Article 1, defines it as: "a change of climate which is attributed directly or indirectly to human activity that alters the composition of the global atmosphere and which is in addition to natural climate variability observed over comparable time periods." This Atlas generally follows the UNFCCC's distinction between "climate change" attributable to human activities altering the atmospheric composition, and "climate variability" attributable to natural causes. Although often used to mean climate change, global warming is only one aspect of this – the increase in global mean temperature.

Coral bleaching The paling in color of corals resulting from a loss of symbiotic algae, in response to abrupt changes in temperature, salinity, and turbidity.

Cryosphere Component of the climate system consisting of all snow, ice, and permafrost on and beneath the surface of the earth and ocean.

Ecosystem A system of interacting living organisms together with their physical environment, which can range from very small areas to the entire Earth.

Emissions Reduction Units Equal to 1 **tonne** of carbon dioxide emissions reduced or sequestered arising from a Joint Implementation (defined in Article 6 of the Kyoto Protocol) project calculated using the **Global warming potential**.

Fugitive emissions Intentional or unintentional releases of gases from anthropogenic activities such as the processing, transmission or transportation of gas or petroleum.

Geothermal Literal meaning: "Earth" plus "heat". To produce electric power from geothermal resources, underground reservoirs of steam or hot water are tapped by wells and the steam rotates turbines that generate electricity.

Glacier A mass of land ice flowing downhill. A glacier is maintained by accumulation of snow at high altitudes, balanced by melting at low altitudes or discharge into the sea.

Global warming Increase in global mean temperature.

Global warming potential (GWP) An index, describing the radiative characteristics of well-mixed greenhouse gases, that represents the combined effect of the differing times these gases remain in the atmosphere and their relative effectiveness in absorbing outgoing long-wave radiation. The GWP of carbon dioxide is 1.

Greenhouse gas Gases in the atmosphere, both natural and anthropogenic, that absorb and emit radiation at specific wavelengths within the spectrum of long-wave radiation emitted by the Earth's surface, the atmosphere, and clouds. This property causes the greenhouse effect. Water vapor (H_2O), carbon dioxide (CO_2), nitrous oxide (N_2O), methane (CH_4), and ozone (O_3) are the primary greenhouse gases in the Earth's atmosphere, but there are a number of entirely human-made greenhouse gases, such as the halocarbons and other chlorine- and bromine-containing substances, dealt with under the Montreal Protocol.

Ice sheet A mass of land ice that is sufficiently deep to cover most of the underlying bedrock topography. There are only two large ice sheets in the modern world, on Greenland and Antarctica.

Ice shelf A floating ice sheet of considerable thickness attached to a coast (usually of great horizontal extent with a level or gently undulating surface); often a seaward extension of ice sheets.

ICT Information computer technology

IPCC Intergovernmental Panel on Climate Change. Established by the World Meteorological Organization (WMO) and the United Nations Environment Programme (UNEP) in 1988 to assess scientific, technical and socio-economic information relevant to the understanding of climate change, its potential impacts and options for adaptation and mitigation. Its Fourth Assessment Report was released during 2007, with presentations of the results of the three working groups involved in assessing the scientific basis, the impacts, adaptations and vulnerabilities, and the mitigation of climate change in the early part of the year, and the synthesis report was scheduled for publication in November 2007.

JI Joint Implementation, a market-based implementation mechanism defined in Article 6 of the Kyoto Protocol, allowing Annex I countries or companies from these countries to implement projects jointly that limit or reduce emissions, or enhance sinks, and to share the Emissions Reduction Units. JI activity is also permitted in Article 4.2(a) of the UNFCCC.

Kyoto Protocol to the United Nations Framework Convention on Climate Change (UNFCCC), adopted at the Third Session of the Conference of the Parties to the UNFCCC in 1997 in Kyoto, Japan. It contains legally binding commitments, in addition to those included in the UNFCCC. Countries included in Annex B of the Protocol (most countries in the Organisation for Economic Cooperation and Development, and countries with economies in transition) agreed to reduce their anthropogenic **greenhouse gas emissions** by at least 5% below 1990 levels in the commitment period 2008 to 2012. The Kyoto Protocol entered into force in 2005 after Annex I countries representing at least 55 percent of emissions by industrialized countries ratified it.

Long-wave radiation Radiation emitted by the Earth's surface, the atmosphere, and clouds, also known as terrestrial or infrared radiation.

Mitigation Anthropogenic intervention to reduce the sources or enhance the sinks of greenhouse gases.

Paleoclimate Climate for periods prior to the development of measuring instruments, including historic and geologic time, for which only proxy climate records are available.

Photovoltaics Panels used to convert the sun's radiation into electricity, also called solar cells.

Precipitation Water in solid or liquid form that falls to Earth's surface from clouds.

Radiative forcing Change in the net vertical radiation at the boundary between the lower and upper atmosphere (the tropopause) due to an internal change or a change in the external forcing of the climate system, such as a change in the concentration of carbon dioxide or the output of the sun.

Terrestrial Relating to the land.

Thermohaline circulation Large-scale density-driven circulation in the ocean, caused by differences in temperature (thermo) and salinity (haline).

UNFCCC United Nations Framework Convention on Climate Change, adopted on 9 May 1992 in New York and signed at the 1992 Earth Summit in Rio de Janeiro by more than 150 countries and the European Community. Its ultimate objective is the "stabilization of greenhouse gas concentrations in the atmosphere at a level that would prevent dangerous anthropogenic interference with the climate system." It contains commitments for all Parties and entered into force in March 1994.

Weather The state of the atmosphere at some place and time described in terms of such variables as temperature, cloudiness, precipitation and wind.

Regions

Annex B countries are listed in Annex B in the **Kyoto Protocol** and have agreed to a target for their greenhouse gas emissions. They include the **Annex I countries** except Turkey and Belarus.

Annex I countries Group of countries included in Annex I (as amended in 1998) to the **UNFCCC**, including all the developed countries in the **OECD**, and economies in transition. By default, the other countries are referred to as non-Annex I countries. Under Articles 4.2(a) and 4.2(b) of the Convention, Annex I countries commit themselves to the aim of returning to their 1990 levels of greenhouse gas emissions by the year 2000.

CIS Commonwealth of Independent States. The states of the former Soviet Union (excluding Estonia, Latvia, Lithuania).

EIT Economies in transition, countries with national economies in the process of changing from a planned economic system to a market economy. Refers to the former communist countries of Europe.

EU European Union. Data for 2000 to 2004 refer to 15 members: Austria, Belgium, Denmark, Finland, France, Germany, Greece, Ireland, Italy, Luxembourg, Netherlands Portugal, Spain, Sweden and United Kingdom. Data for 2004 onwards refer to 25 member countries, which includes Cyprus, the Czech Republic, Estonia, Hungary, Latvia, Lithuania, Malta, Poland, Slovakia and Slovenia.

G8 countries Group of Eight: Canada, France, Germany, Italy, Japan, the UK, USA, and Russia. The heads of government hold an annual economic and political summit meeting in the country currently holding the rotating presidency.

LDCs The least developed countries, identified as such by the United Nations.

OECD Organisation for Economic Co-operation and Development. 30 member countries sharing a commitment to democratic government and the market economy.

OPEC Organization of the Petroleum Exporting Countries.

Impacts, vulnerability and adaptation

Climate change impacts are the consequences of natural and human systems. The impacts depend on the **vulnerability** of the system, in the climate change literature defined as a function of the character, magnitude, and rate of climate variation to which a system is exposed, its sensitivity, and its adaptive capacity. However, vulnerability has other common definitions. In disaster planning, risk is the outcome of vulnerability (social, economic and environmental exposure and sensitivity) and hazard (the probability and magnitude of an extreme event). In development planning and poverty assessment, vulnerability is described as exposure to multiple stresses, to shocks and to risk over a longer time period, with a sense of defencelessness and insecurity.

Both impacts and vulnerability may be reduced by **adaptation** – adjustments in natural or human systems to a new or changing environment. Various types of adaptation can be distinguished, including anticipatory and reactive adaptation, private and public adaptation, and autonomous and planned adaptation. For people, adaptation can be seen as a process of social learning. **Adaptive capacity** is the ability to understand climate changes and hazards, to evaluate their consequences for vulnerable peoples, places and economies, and to moderate potential damages to take advantage of opportunities, or to cope with the consequences.

Theory, prediction, forecasts and scenarios

A scientific **theory** is a coherent understanding of some aspect of our world, based on a well-established body of observations and interpretation. In popular culture, the label "theory" may be used in a derogatory manner to refer to propositions put forward to challenge a mainstream view.

There is some confusion regarding how we view the future. In experimental language, we talk of predicted outcomes. So, a computer simulation of climate change predicts global warming, say of 3°C by 2100. This prediction depends on a set of underlying assumptions.

A climate *prediction* is usually in the form of probabilities of climate variables such as **temperature** or **precipitation**, with lead times up to several seasons. The term climate *projection* is commonly used for longer-range predictions that have a higher degree of uncertainty and a lesser degree of specificity. For example, this term is often used to describe future **climate change**, which depends on the uncertain consequences of greenhouse gas emissions and land use change, in addition to the feedbacks within the atmosphere, oceans and land surface.

It is extremely difficult to anticipate future greenhouse gas emissions. Countries may adopt stronger controls, industry and technology might reduce their carbon intensity, or consumers might rebel and ecological feedbacks might accelerate climate change. Where our understanding of the future is weak, we often use the term **scenario**: a plausible and often simplified description of how the future may develop, based on a coherent and internally consistent set of assumptions about key driving forces and relationships. Scenarios are neither predictions nor forecasts and sometimes may be based on a narrative storyline.

Greenhouse gas emission scenarios

To enable comparisons to be made between predictions, in 1996 the IPCC issued its *Special Report on Emissions Scenarios* (SRES), which outlined a number of possible scenarios. These are now used by scientists to clarify the assumptions behind different emission pathways and the consequences for climate change.

Scenario A1 represents a future world of very rapid economic growth, low population growth, and rapid introduction of more efficient technologies. Major underlying themes are economic and cultural convergence and capacity building, with a substantial reduction in regional differences in per capita income. The A1 scenario family develops into three alternative directions of technological change in the energy system: fossil-intensive (A1FI), non-fossil energy sources (A1T), or a balance across all sources (A1B).

Scenario A2 portrays a very heterogeneous world. The underlying theme is that of strengthening regional cultural identities, with high population growth rates, and less concern for rapid economic development.

Scenario B1 represents a convergent world with a global population that peaks in mid-century, rapid change in economic structures toward a service and information economy, with reductions in material intensity, and the introduction of clean and resource-efficient technologies. The emphasis is on global solutions to economic, social, and environmental sustainability, including improved equity, but without additional climate initiatives.

Scenario B2 depicts a world in which the emphasis is on local solutions to economic, social, and environmental sustainability. It is a heterogeneous world with less rapid, and more diverse technological change than in A1 and B1.

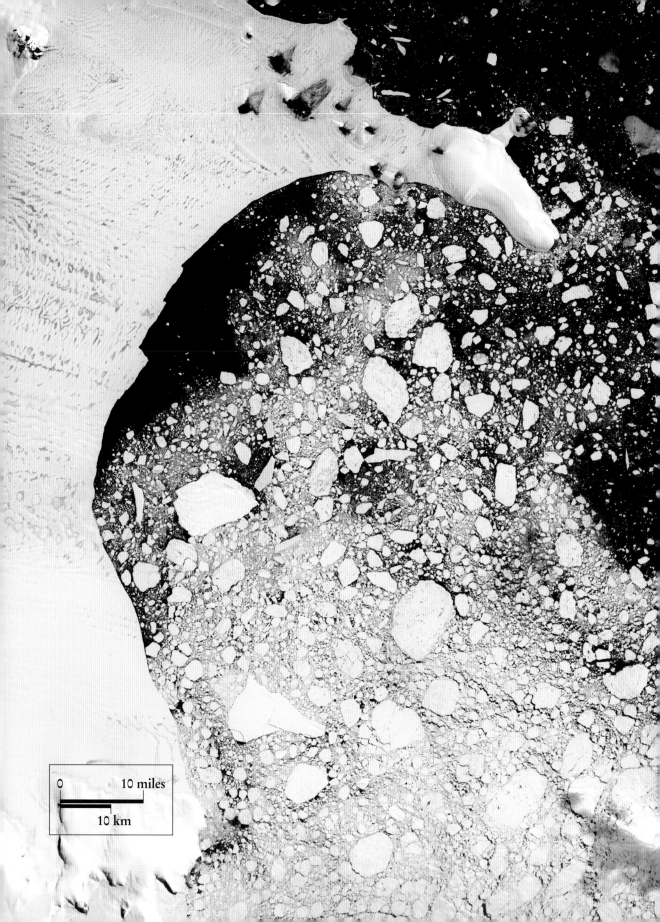

PART 1 SIGNS OF CHANGE

Hindsight is an amazing thing. It allows all of us to say that someone should have recognized the signs, have seen it coming. Awareness of our present conundrums is much more difficult, especially when some of the future consequences are hard to imagine.

Climate change is especially challenging. Most of us have not seen the early signs ourselves. Scientists working in the Mauna Loa observatory in Hawaii began monitoring carbon dioxide levels in the 1950s and saw the concentration increase year by year, but this was not considered generally newsworthy. Over the last decade, people in northern regions have seen grasses sprouting where none had been able to grow before, and watched their roads buckle and the foundations of their homes shift as the permafrost melted beneath them. Gardeners and birders with a sharp eye and long commitment have seen a shift in the timing of blossoms and the length of growing seasons, as well as the arrival, departure, and nesting time of local birds. But to recognize these signs as pieces of a global pattern requires assembling a vast array of observations and knowledge.

Since 1990 scientists have made an increasingly elaborate effort to collect and analyze the pieces of this larger picture. In this section of the atlas, we report the findings of global monitoring systems of the oceans, atmosphere and cryosphere, over tropical and temperate lands, in ecosystems from coral reefs to mountains, glaciers and polar ice sheets. These are just a few examples from among the hundreds of studies investigating possible changes related to warmer temperatures and other climatic changes.

There will always be some uncertainty about a system as complex as the world's climate and diverse as the world's ecosystems. Our understanding of some impacts is still developing. However, this weight of evidence raises the question of another type of uncertainty: when will governments, corporations, and individuals progress from tentative, limited responses to recognizing the long-term import of these warning signs as harbingers of our future?

"Amplification of the Earth's natural greenhouse effect by the buildup of various gases introduced by human activity has the potential to produce dramatic changes in climate. Only by taking action now can we ensure that future generations will not be put at risk."

49 Nobel Prize winners and 700 members of the National Academy of Sciences, 1990

SIGNS OF CHANGE

Trends and events consistent with theories
of climate change
1990s–April 2007

Changes resulting from:

local temperature rise

extreme heat and/or drought

extreme precipitation and/or wind

changing animal and plant behaviour

Increases in global average air and ocean temperatures, widespread melting of snow and ice, and rising sea levels led the Intergovernmental Panel on Climate Change (IPCC) to report, in February 2007, that "warming of the climate system is unequivocal".

The world is experiencing increasingly uncommon weather, and the implications for day-to-day life are becoming more apparent. Naturalists' observations of animal and plant behaviour suggest that ecosystems are already being forced to adjust. In April 2007, the IPCC stated with "high confidence" that recent warming has affected terrestrial, marine and freshwater biological systems, glaciers and rivers. Based on an analysis of over 29,000 data sets, contained in 75 studies from around the world, it concluded that over 90 percent of the observed changes were consistent with climate change.

A single extreme weather event or change in the natural environment does not prove that humans are changing the climate. However, the proven physical science, the history of recent observations, and the consistency in model assessments all support only one explanation: the emission of greenhouse gases by human activity is causing profound changes to the climate system and to the world we live in.

Canadian polar bears

Arctic sea ice is declining by around 8% each decade, reducing polar bears' hunting season and resulting in poorer health and reproductive success.

Alaskan permafrost

A temperature increase of 3°–4°C since 1950 is having a visible effect in Alaska. Roads and buildings in some areas are subsiding as the permafrost melts, and the absence of summer sea-ice is leading to coastal erosion, making some low-lying communities unviable.

Washington, DC flowering

Trees are flowering an average of 4.5 days earlier in Washington, DC.

Mosquitoes

Scientists have detected a genetic adaptation to warmer temperatures in a type of North American mosquito. It is entering its winter dormancy 9 days later than it did 30 years ago, prolonging the period during which it can spread disease.

Atlantic hurricanes

The 2005 Atlantic hurricane season broke records for the frequency of storms, and for the number of category 5 hurricanes. Whether these are signs of climate change or not is a fiercely debated topic.

Tropical Andes

There has been a widespread retreat of mountain glaciers during the 20th century.

Floods in Bolivia

Heavy rain and flooding early in 2006 affected around 17,500 people.

South Atlantic hurricane

The first hurricane ever observed in the South Atlantic hit Brazil in 2004.

Floods in Brazil

Tens of thousands of people were left homeless and over 160 killed in 2004 by mudslides and floods.

Drought in southern Brazil

Drought conditions in 2006 led to a 10% decrease in soybean yield.

Larsen B ice shelf

In 2002, 1,250 square miles (3,250 sq km) of ice shelf broke away from the Antarctic Peninsula. This was followed by an unexpectedly rapid increase in the rate of glacial flow and ice sheet retreat.

Warning Signs

New records and observations around the world are consistent with scientists' expectations of climate change.

European autumn

In many countries north of the Alps, autumn 2006 was the warmest since official measurements began. Records for central England date back to 1659.

European heatwave

About 35,000 deaths were related to heat in France, Italy, Netherlands, Portugal, Spain and the UK in 2003.

European Alps

The average loss of thickness of glaciers in 2003 was nearly twice that of the previous record year of 1998.

European butterfly ranges

Of 35 European non-migratory butterfly species studied, 22 shifted their ranges north by 20–150 miles (35–240 km) during the 20th century, and only one shifted south.

Spanish drought

In 2005 Spain experienced the driest winter and early spring since records began in 1947.

Siberian melt

Average temperatures in the west have risen by 3°C since the 1960s, leading to melting of the permafrost.

Changes in China

Increasing temperatures, rainfall, and glacial melt rates are consistent with expectations of climate change in northwest China.

Indian heatwave

More than 1,500 fatalities in India and Pakistan in 2003 were caused by temperatures over 50°C.

Drought in Horn of Africa

In 2006 at least 11 million people were affected by food shortages caused by drought in Burundi, Djibouti, Eritrea, Ethiopia, Kenya, Somalia and Tanzania.

Floods in Horn of Africa

In 2006 the region experienced the worst floods in 50 years. They claimed over 600 lives in Ethiopia, and affected hundreds of thousands of people in Somalia.

Drought in China

In 2006, severe drought in the north led to 12% of the nation's agriculture being affected. Severe drought in the south affected 18 million people.

Floods in China

Heavy rain led to severe flooding and landslides in June 2006, affecting 17 million people in southern China.

Asian summer monsoon

Heavy rain and flooding in parts of northern India, Nepal and Bangladesh in 2004 left 1,800 dead and millions stranded.

Coral bleaching

In 1998, the Great Barrier Reef and coral reefs elsewhere experienced the most severe bleaching ever recorded.

Drought in Australia

Abnormally low or absent rainfall in the decade up to 2007 severely affected cattle farmers, and led to cities imposing water restrictions. Arable farmers were warned, in April 2007, of the possibility of restrictions on water for irrigation.

Floods in New Zealand

In February 2004 severe floods caused loss of life and extensive damage to property and agricultural land.

Average temperatures in the Arctic and much of the Antarctic are increasing, and ice is melting in both regions.

Floating Arctic ice has covered the North Pole for millions of years. Its extent fluctuates with the seasons, but has declined by 15 to 20 percent over the past 30 years. Already, the North Pole is free of ice in some summers and by the end of this century the ice may largely disappear for entire summers. The huge landmass of Greenland is covered with an ice sheet 3.2 kilometres deep in places. Warmer air is causing this to melt at an alarming rate.

The Antarctic is also changing, although not uniformly. The Antarctic Peninsula has experienced marked warming, with visible changes occurring during the working lives of the scientists studying its flora and fauna. Warmer ocean waters are thinning and weakening the ice shelves, and warmer summers are producing melt water that floods crevasses, creating a wedge more dense than the surrounding ice that drives through to the shelf base.

In East Antarctica, although glacier movement appears to be accelerating, the changes are much less dramatic than those on the Peninsula. In West Antarctica, however, a coastal section of the ice sheet is now thinning quite rapidly.

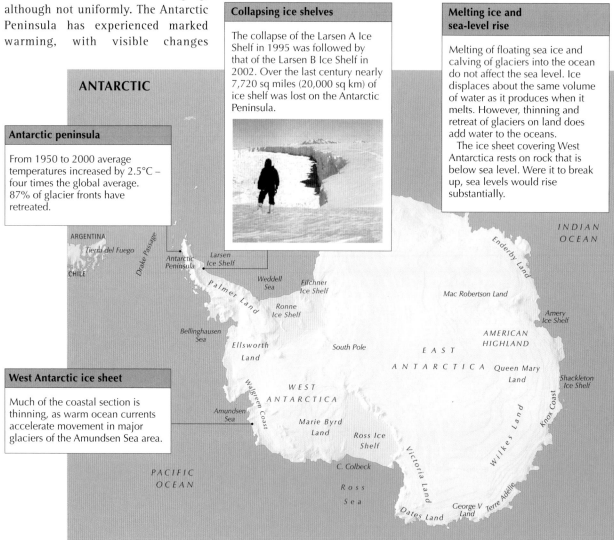

ANTARCTIC

Antarctic peninsula

From 1950 to 2000 average temperatures increased by 2.5°C – four times the global average. 87% of glacier fronts have retreated.

Collapsing ice shelves

The collapse of the Larsen A Ice Shelf in 1995 was followed by that of the Larsen B Ice Shelf in 2002. Over the last century nearly 7,720 sq miles (20,000 sq km) of ice shelf was lost on the Antarctic Peninsula.

Melting ice and sea-level rise

Melting of floating sea ice and calving of glaciers into the ocean do not affect the sea level. Ice displaces about the same volume of water as it produces when it melts. However, thinning and retreat of glaciers on land does add water to the oceans.

The ice sheet covering West Antarctica rests on rock that is below sea level. Were it to break up, sea levels would rise substantially.

West Antarctic ice sheet

Much of the coastal section is thinning, as warm ocean currents accelerate movement in major glaciers of the Amundsen Sea area.

ARGENTINA
Tierra del Fuego
Drake Passage
CHILE
Antarctic Peninsula
Larsen Ice Shelf
Palmer Land
Weddell Sea
Filchner Ice Shelf
Ronne Ice Shelf
Bellinghausen Sea
Ellsworth Land
South Pole
EAST ANTARCTICA
Mac Robertson Land
Enderby Land
INDIAN OCEAN
Amery Ice Shelf
AMERICAN HIGHLAND
Queen Mary Land
Shackleton Ice Shelf
Amundsen Sea
Walgreen Coast
WEST ANTARCTICA
Marie Byrd Land
Ross Ice Shelf
Knox Coast
Wilkes Land
PACIFIC OCEAN
C. Colbeck
Ross Sea
Victoria Land
Oates Land
George V Land
Terre Adélie

Polar Changes

Warming in the polar regions is driving large-scale melting of ice that will have both local and global consequences.

ARCTIC

The extent of the Arctic sea ice has decreased by 14% since the 1970s. In August 2005 the ice cap shrank to its smallest recorded extent. By 2080 sea ice is expected to disappear in the summer months.

While an open Arctic sea would have benefits for shipping, fishing and the exploitation of minerals, it would be at great cost to the environment and to traditional livelihoods. A delay in the formation of the winter ice, an earlier break-up of ice in the spring, and thinner ice year round makes it hard for indigenous people using largely traditional methods to make a living.

The Alaskan permafrost is also melting, weakening the coastline and creating subsidence that is causing roads and buildings to collapse. Around 15% of the Arctic tundra has already thawed.

Extent of summer ice 1979

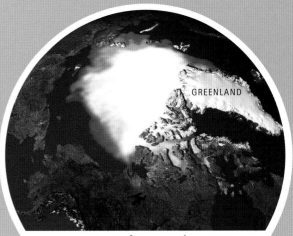

Extent of summer ice 2005

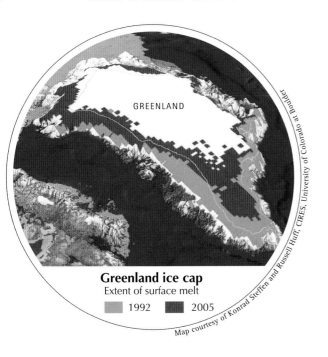

Greenland ice cap
Extent of surface melt

■ 1992   ■ 2005

Map courtesy of Konrad Steffen and Russell Huff, CIRES, University of Colorado at Boulder

Each summer, parts of the Greenland ice sheet melt, at the edges and on the surface. Although the melt area varies each year, the overall trend since 1979 has been upwards. Surface melt water finds its way through crevasses to the base of the ice, and forms a thin film between ice and bedrock. There are fears that this could increase the speed at which the ice sheet slides towards the sea.

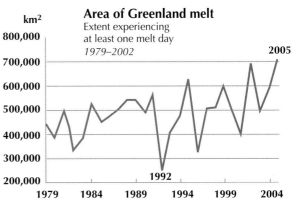

Area of Greenland melt
Extent experiencing at least one melt day
1979–2002

km²

800,000	
700,000	2005
600,000	
500,000	
400,000	
300,000	
200,000	1992

1979   1984   1989   1994   1999   2004

GLACIAL CHANGE

since 1950s
selected glaciers

Around the world, the fronts of most glaciers are receding to higher altitudes, and at such a rate that glaciologists, mountaineers, tourists and local residents are astonished by the changes that have occurred in just a decade or two.

Because glaciers respond to temperature trends, rather than single warm years, they provide valuable evidence of long-term change. And because the tell-tale signs of their expansion and retraction are clearly visible, scientists are able to draw conclusions about climate change from periods well before instrumental records became widespread. Tree stumps and even, in 1991, human remains that have been preserved in the ice for thousands of years, are now being revealed.

Warmer temperatures lead to increased precipitation, so in some parts of the world, even though glaciers are retreating they are getting thicker. Overall, however, the total volume of ice held in glaciers is decreasing dramatically.

Glacial melting changes the flow of rivers, adding to water stress for millions of people. Lakes formed from melting glaciers are unstable, prone to abrupt collapse and flooding, threatening lives downstream.

Alaska
Glaciers are both retreating and thinning.

Greenland
A rapid retreat and loss of ice mass in Greenland is giving cause for concern.

Canadian Rockies
Tree stumps are being exposed for the first time in 2,500 years as glaciers recede.

Popocatepetl
The Ventorrillo glacier showed signs of retreat between 1950 and 1982.

Northern Andes
The Quelccaya Glacier, Peru, is retreating 10 times more rapidly than it did in the 1970s and 1980s – by up to 30 meters a year.

USA
The South Cascade glacier, Washington, and the **Arapaho glacier**, Colorado, are retreating. Since 1960 the Arapaho glacier has also thinned by 40 meters.

1898

2003

Southern Andes
About half the glaciers surveyed in Chile show signs of retreat.

Antarctic Peninsula
85% of glaciers are retreating.

Glacial Retreat

Most of the world's glaciers are retreating at unprecedented rates – a clear sign of warming.

Scandinavia

Many glaciers are retreating, although increased snowfall is adding to their mass.

Tien Shan

The 400 glaciers in the north of the range have lost 25% of their volume since 1955.

Himalayan and other Asian glaciers

Almost all glaciers surveyed are in retreat.

Irian Jaya

The Meren glacier had disappeared by 2000. The Carstenz and Northwall Firn glaciers have lost 20% of their area since 2000.

Europe in Africa

Glaciers have shrunk to a third of their 1850 extent, and have lost half their volume.

New Zealand

Three-quarters of glaciers studied showed signs of retreat.

East Africa

The famous snow-capped peaks on Mounts Kenya and Kilimanjaro are shrinking so rapidly that they may vanish by 2025. The Rwenzori Mountains are also melting rapidly.

Glacial lakes

Lakes have developed at the foot of glaciers in Bhutan in the Himalayas from the increasing volume of melt water. The natural dams holding the water back may break, unleashing torrents of water and threatening the lives and livelihoods of the people living in valleys below.

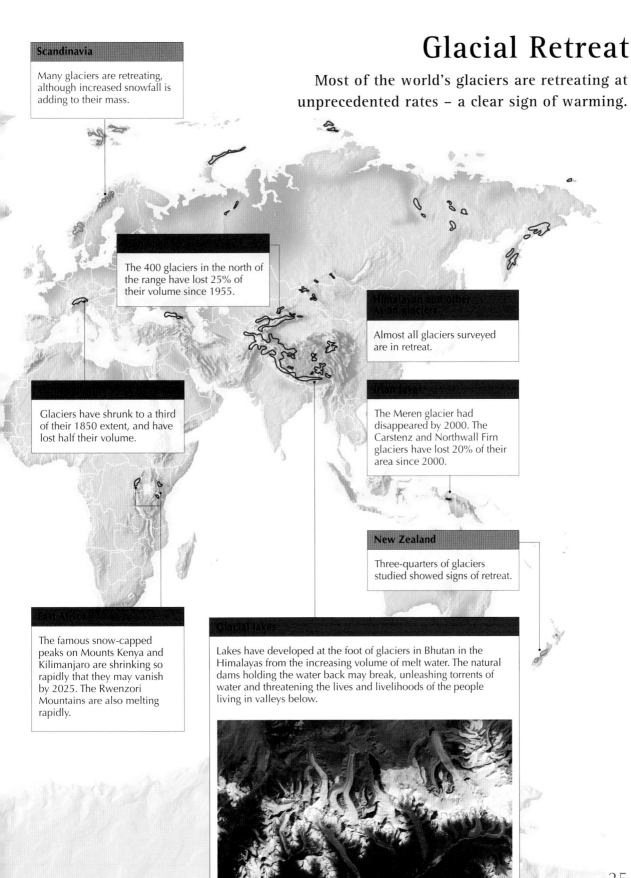

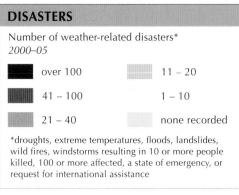

Extreme weather events have, in recent years, caused heavy loss of life, and extensive damage. In Mumbai, India, 800 people were killed in July 2005 as floods swept through the streets after nearly a meter of rain fell in just 24 hours. Repeated flooding on the Danube River will add to the high damage costs incurred in recent years.

It is difficult to link any single example of the dramatic power of the Earth's meteorological system to climate change. It is, however, possible to collect data on the frequency and intensity of disasters – events that overwhelm response capacities – and to track changes over time. Records kept by the International Disaster Database indicate that the incidence of flood and windstorm disasters has not only increased markedly since the 1960s, but the events themselves are more intensive, last longer and affect more people.

Most climate models indicate an increase in the frequency and duration of extreme events in some areas. For instance, in the USA total precipitation has increased between 5 and 10 percent over the last decade, and rain and snow are falling in fewer, more extreme, events. Summer heatwaves are likely to become more common in Europe, and by 2030 most of Australia is expected to experience between 10 and 50 percent more days over 35°C.

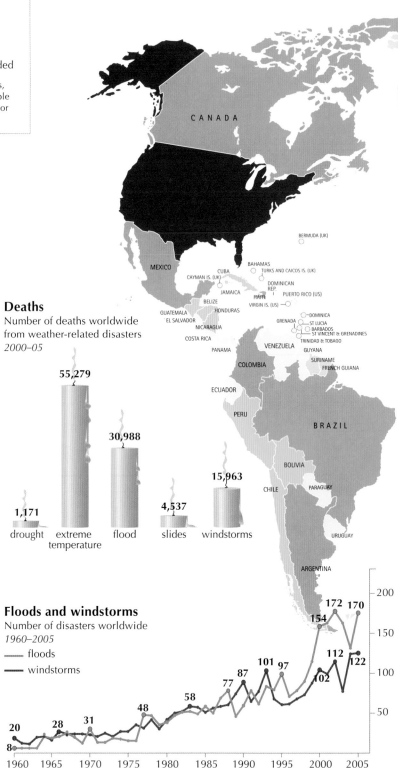

Deaths

Number of deaths worldwide
from weather-related disasters
2000–05

- drought: 1,171
- extreme temperature: 55,279
- flood: 30,988
- slides: 4,537
- windstorms: 15,963

Floods and windstorms

Number of disasters worldwide
1960–2005

- floods
- windstorms

8, 20, 28, 31, 48, 58, 77, 87, 101, 97, 154, 172, 170, 112, 102, 122

1960 1965 1970 1975 1980 1985 1990 1995 2000 2005

200 — 150 — 100 — 50

Everyday Extremes

Weather-related disasters are becoming increasingly common around the world.

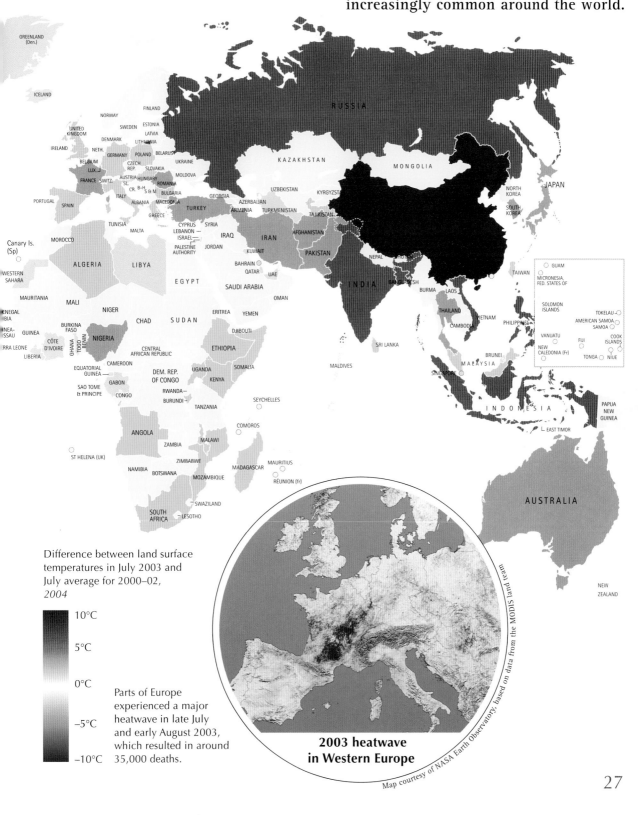

Difference between land surface temperatures in July 2003 and July average for 2000–02, *2004*

10°C
5°C
0°C
−5°C
−10°C

Parts of Europe experienced a major heatwave in late July and early August 2003, which resulted in around 35,000 deaths.

2003 heatwave in Western Europe

Map courtesy of NASA Earth Observatory, based on data from the MODIS land team

PART 2 FORCING CHANGE

The evidence for climate change presented in this atlas is drawn from three strands. The first, the physical understanding of the effect of increasing greenhouse gas concentrations on radiation and the Earth's climate system, has been documented for over a century. Imagine the atmosphere as a thermal blanket such as a comforter or duvet that keeps our sleeping bodies warm by absorbing the heat radiating from us. If we fill the holes in the duvet with extra feathers, it will keep us even warmer. This is what we are doing to the Earth's atmosphere – adding greenhouse gases and keeping more heat in the atmosphere.

The second set of evidence is the enormous volume of data. The global temperature is rising, and scientific observations from around the world are showing changes consistent with scientists' expectations. In 1995, the Intergovernmental Panel on Climate Change, in its second periodic review of climate change science, concluded that the evidence suggests a discernible human influence on global climate. Now, a decade later, that human influence is widely acknowledged. Most importantly, we are seeing the evidence of global warming in countless climate records and changes in ecosystems and species.

The third strand of evidence brings together the physical understanding and observations in the modelling of the atmosphere and oceans, using the most advanced science and computers. Models are tested for how well they represent the known history of climate changes. If the model passes this test, it is then run forward to project what climate change might look like in the remainder of this century and beyond. All of the models show global warming in response to increasing greenhouse gas concentrations. Natural factors alone cannot explain the observed global warming over the last few decades or so; in order to match the observations, it is necessary to include the role of greenhouse gases in trapping increased radiation.

Confidence in this outline of climate change – the pattern of global warming – has not come easily. While huge investments in climate science have been made, with new monitoring systems in the oceans and space, using state-of-the-art computers, and mobilizing thousands of experts around the world, our understanding remains incomplete. Scientists continue to observe, analyze, debate, and refine areas of understanding, albeit with much more media coverage.

Although we are still learning, the essential facts – the three strands of evidence – are solid. The quickening in the pace of observed climate change has been felt by all who study this literature, and is now widely reported on the news. Among the uncertainties, we recognize many ways in which the threats may not be as large as sometimes portrayed. Equally, there is evidence of risks significantly in excess of any we would wish to experience ourselves.

"Most of the observed increase in globally averaged temperatures since the mid-20th century is *very likely* due to the observed increase in anthropogenic greenhouse gas concentrations."

IPCC Working Group I, 2007

Without the natural greenhouse effect, which captures and holds some of the sun's heat, humans and most other life-forms would not have evolved on Earth. The average temperature would be −18°C, rather than 15°C. But the enhancement of the greenhouse effect in recent years is driving increases in temperature and many other associated changes in climate.

Solar radiation passes through the atmosphere, and heats the surface of the Earth. Some of that energy returns to the atmosphere, but not all of it makes it through the layer of gases that covers the Earth like the glass of a greenhouse. It is this layer of gases that is central to the climate change currently being experienced.

Over the last 250 years or so human activity, such as the burning of fossil fuels, the removal of forests that would otherwise absorb carbon dioxide, and their replacement with intensive livestock ranching, has released a range of "greenhouse gases" into the atmosphere; other industrial pollutants have reacted to form tropospheric ozone. The capacity of the atmosphere to absorb heat and emit it back to Earth has been substantially increased.

Some greenhouse gases only stay in the atmosphere for a few days, but others remain for decades, centuries, or millennia. Greenhouse gases emitted today will drive climate change well into the future and the process cannot be quickly reversed.

Warming leads to feedbacks that accelerate the greenhouse effect. Higher temperatures mean more evaporation and water vapor in the atmosphere, which also absorbs heat. Snow cover is likely to decrease, allowing more solar radiation to reach the ground. Warming also thaws the permafrost, which might result in further releases of greenhouse gases, including methane.

THE GREENHOUSE EFFECT

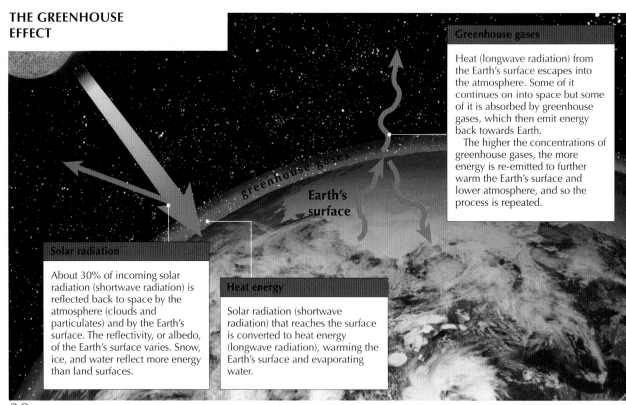

Greenhouse gases

Heat (longwave radiation) from the Earth's surface escapes into the atmosphere. Some of it continues on into space but some of it is absorbed by greenhouse gases, which then emit energy back towards Earth.

The higher the concentrations of greenhouse gases, the more energy is re-emitted to further warm the Earth's surface and lower atmosphere, and so the process is repeated.

greenhouse gases

Earth's surface

Solar radiation

About 30% of incoming solar radiation (shortwave radiation) is reflected back to space by the atmosphere (clouds and particulates) and by the Earth's surface. The reflectivity, or albedo, of the Earth's surface varies. Snow, ice, and water reflect more energy than land surfaces.

Heat energy

Solar radiation (shortwave radiation) that reaches the surface is converted to heat energy (longwave radiation), warming the Earth's surface and evaporating water.

The Greenhouse Effect
The increasing concentration of greenhouse gases is trapping more heat.

Change in atmospheric composition since the Industrial Revolution
Northern hemisphere
pre-1750 and 2003
parts per million (ppm) – CO_2
parts per billion (ppb) – N_2O, CH_4, O_3
parts per trillion (ppt) – CCl_4

CCl_4, a solvent used in drycleaning, is a halocarbon, a group of climate-forcing gases that includes CFCs, and their replacement compounds HCFCs and HFCs.

pre-1750   2003

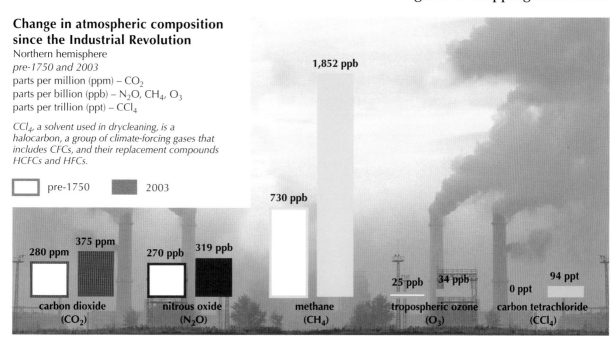

1,852 ppb
730 ppb
280 ppm   375 ppm
270 ppb   319 ppb
25 ppb   34 ppb
0 ppt   94 ppt

carbon dioxide
(CO_2)

nitrous oxide
(N_2O)

methane
(CH_4)

tropospheric ozone
(O_3)

carbon tetrachloride
(CCl_4)

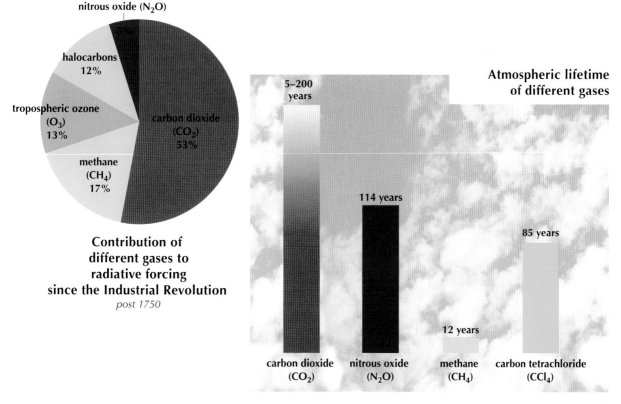

nitrous oxide (N_2O)
5%
halocarbons
12%
tropospheric ozone
(O_3)
13%
carbon dioxide
(CO_2)
53%
methane
(CH_4)
17%

Contribution of different gases to radiative forcing since the Industrial Revolution
post 1750

Atmospheric lifetime of different gases

5–200
years
114 years
85 years
12 years

carbon dioxide
(CO_2)

nitrous oxide
(N_2O)

methane
(CH_4)

carbon tetrachloride
(CCl_4)

By capturing more energy in the atmosphere and the Earth's surface, the greenhouse effect is expected to alter the climate system. This giant "heat-distribution engine" relies primarily on atmospheric and oceanic circulation to move heat energy and distribute it more evenly around the world. The more heat energy there is, the harder it works.

Most solar radiation and heating of the surface occurs at the equator, where the sun's rays are nearly perpendicular to the surface all year round. The poles receive much less radiation because of the Earth's orbit and tilt relative to the sun. The climate system redistributes this heat energy more equally.

Atmospheric and oceanic circulation contribute equally in moving energy from the equator towards the poles. An increase in the difference between the temperature in the tropics and at the poles could disrupt the climate in many ways.

The process of heat energy distribution by the global climate system is largely responsible for regional climates. Changes to the global system mean inevitable changes in the climatic conditions we have come to expect. Warmer summers, heat waves, drier winters, less snowfall, and changing frequency and intensity of storms, are all possible results.

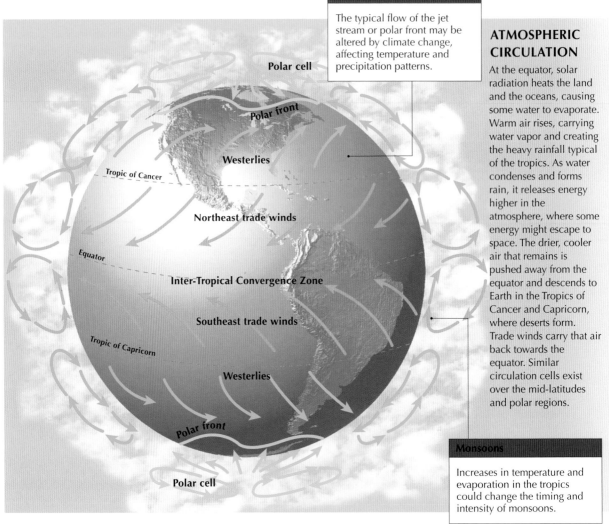

Jet stream

The typical flow of the jet stream or polar front may be altered by climate change, affecting temperature and precipitation patterns.

Polar cell

Polar front

Westerlies

Tropic of Cancer

Northeast trade winds

Equator

Inter-Tropical Convergence Zone

Southeast trade winds

Tropic of Capricorn

Westerlies

Polar front

Polar cell

ATMOSPHERIC CIRCULATION

At the equator, solar radiation heats the land and the oceans, causing some water to evaporate. Warm air rises, carrying water vapor and creating the heavy rainfall typical of the tropics. As water condenses and forms rain, it releases energy higher in the atmosphere, where some energy might escape to space. The drier, cooler air that remains is pushed away from the equator and descends to Earth in the Tropics of Cancer and Capricorn, where deserts form. Trade winds carry that air back towards the equator. Similar circulation cells exist over the mid-latitudes and polar regions.

Monsoons

Increases in temperature and evaporation in the tropics could change the timing and intensity of monsoons.

The Climate System

The entire climate system is adjusting to an increase in the heat trapped in the Earth's atmosphere.

Ocean conveyor belt

The ocean currents distribute energy from solar heating. Some currents are mainly driven by winds and tides, but others are primarily driven by differences in ocean temperature and salt concentration. Water naturally mixes to try and even out the distribution of heat and salt. Circulation occurs as the denser colder, saltier water drops below the warmer, fresher water. These differences drive the "thermohaline circulation", or what is sometimes called the "ocean conveyor belt".

If climate change warms the polar waters and/or decreases their salinity by adding fresher water from melting glaciers, the difference in water density will decline and the circulation pattern is expected to slow down or even collapse. Scientists call the potential collapse a "low-probability, high-impact" scenario. Many think it is very unlikely in this century. But, if the circulation were to collapse, it could happen within two decades.

Even a slowing of the circulation is likely to have a wide-ranging impact. Descending cold waters carry carbon dioxide into deep water, away from the atmosphere. Elsewhere, parts of the world rely on upwelling waters to bring nutrients from the bottom of the ocean to the surface areas, where they help support ecological communities and fisheries.

Storms

Increased temperatures and water vapor in the air over tropical oceans will create improved spawning conditions for cyclones, hurricanes and typhoons.

Northern Europe

Europe has a mean annual surface air temperature some 5–7°C warmer than regions at the same latitude in the Pacific. A collapse, or even a slowing, of the circulation would reduce the amount of warm, tropical water brought north. Although this cooling might be offset by atmospheric warming, the consequences could be severe.

Research comparing flows in 2004 with those in 1992 indicates a 30% reduction in the cold southerly flow, and a 30% increase in the amount of warm water peeling off from the northerly flow and returning to the tropics.

Europe

The heatwaves in Europe may be linked to this global ocean circulation.

Sahel

Researchers are investigating whether variability in rainfall in the Sahel region is associated with changes in the thermohaline circulation system.

Sea-to-air heat transfer

Atlantic Ocean

Gulf Stream

Indian Ocean

Solar warming of oceans

Pacific Ocean

Gulf Stream

warm shallow current

cold and salty deep current

El Niño

The frequency and intensity of El Niño events, triggered when warm water in the Pacific extends east, may be affected by changes in climate systems. During El Niño years, rainfall follows the warm water, leading to flooding in Peru, drought in Indonesia and Australia, and a disruption to climate patterns elsewhere in the world.

Records of the Earth's past climates, reconstructed from ice cores, paleoclimatic fingerprints in sediments, and even ship records of sea-surface temperatures, confirm that global warming is real and unprecedented. The Earth is warmer than in the past millennium, and the commitment to future warming is evident in the record levels of carbon dioxide (CO_2) in the atmosphere. Modern human societies and economies have never faced such conditions.

A physical explanation of the greenhouse effect was already well developed 100 years ago, and since then scientists have identified trends in atmospheric composition and temperature over hundreds of thousands of years. The concentrations of carbon dioxide and methane in the atmosphere are now at the highest levels for over 650,000 years. The atmospheric concentration of carbon dioxide has risen from 315 parts per million (ppm) in the 1950s to over 380 ppm in 2006. The level of carbon dioxide in the atmosphere is very closely linked to global temperatures.

The physical understanding of the climate system, with its many variables,

is captured in computer models. Their efficacy can be tested against the record of past climates, and many have managed to reconstruct past climates reasonably accurately.

Fluctuations in past temperatures have been shown to be caused by natural forces, such as cycles of solar energy, changes in the Earth's orbit and volcanic eruptions that send gases and dust into the atmosphere. However, the variability and trends in historical global temperatures can only be explained if both natural forces and greenhouse gas emissions from human activity are included in the models. This validation of models of the physics of the climate system gives scientists the confidence to use these models to project future climate change.

> CO_2 is about 30% higher, and methane 130% higher than at any time. And the rates of increase are absolutely exceptional: for CO_2, 200 times faster than at any time in the last 650,000 years.
>
> Thomas Stocker, European Project for Ice Coring in Antarctica

Accumulated knowledge

Total number of scientific articles referring to climate change and published in journals indexed in Web of Science
1971–2005

The science of climate change is well established, with over 2,000 articles now appearing every year in peer-reviewed journals. Over 3,000 scientists in the Intergovernmental Panel on Climate Change (IPCC) collect and review this knowledge.

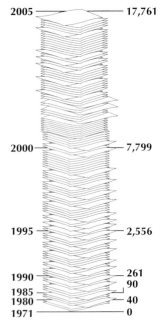

2005	17,761
2000	7,799
1995	2,556
1990	261
1985	90
1980	40
1971	0

CO_2 fluctuations
400,000 years ago to present day

——— concentration in ice-core samples
——— concentration in atmosphere

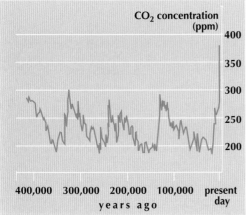

CO_2 concentration (ppm)

400
350
300
250
200

400,000   300,000   200,000   100,000   present day
years ago

Link between CO_2 and temperature
160,000 years ago to present day

——— CO_2 concentration in atmosphere
——— global temperature

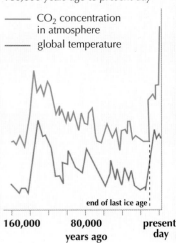

end of last ice age

160,000   80,000   present day
years ago

Interpreting Past Climates

Concentrations of carbon dioxide and methane are higher than they have ever been in the last 650,000 years. The Earth is warmer than at any time in the past 1,000 years.

A millennium of warming in Northern Hemisphere
1000–2000

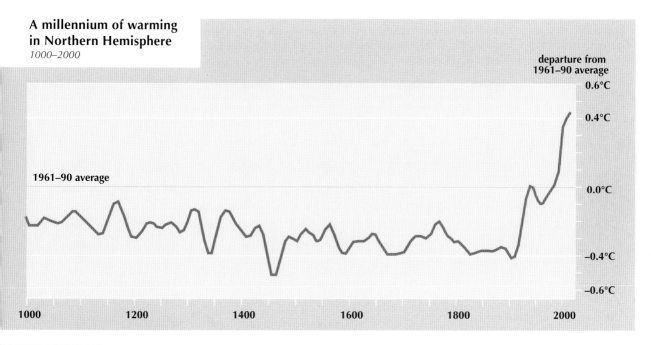

departure from 1961–90 average

0.6°C

0.4°C

1961–90 average

0.0°C

−0.4°C

−0.6°C

1000   1200   1400   1600   1800   2000

If the natural factors that affect the climate are fed into a computer model, the resulting graph does not match the temperature fluctuations that have occurred. The warming since 1960 can only be accounted for if both natural and human-induced greenhouse gases are taken into account.

Accounting for warming
Comparison of temperature rise with results of models taking into account natural and human factors
1850–2000

——— observed global temperature
——— model results

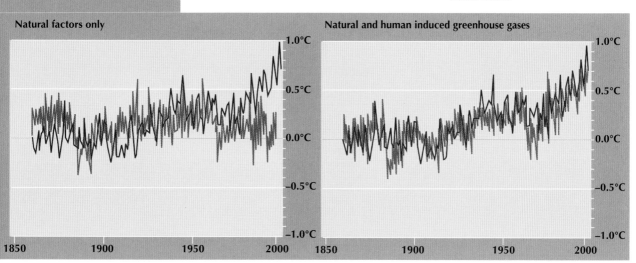

Natural factors only

1.0°C

0.5°C

0.0°C

−0.5°C

−1.0°C

1850   1900   1950   2000

Natural and human induced greenhouse gases

1.0°C

0.5°C

0.0°C

−0.5°C

−1.0°C

1850   1900   1950   2000

Climatologists are confident of future climate change. The same computer models used to reconstruct past climates and to forecast weather are used to project climatic conditions in response to increasing greenhouse gas concentrations.

The starting point for projecting future climates is a series of scenarios of the extent of the use of fossil fuels, the development and distribution of technologies that reduce emissions, and of the speed of population and economic growth. These different pathways of future energy use and development are associated with different rates of greenhouse gas emissions, which, in turn, lead to varying projections of their concentrations in the atmosphere. The pre-industrial concentration of carbon dioxide (CO_2) was 280 parts per million in the atmosphere; even the lowest projected increase is for a concentration of over 520 parts per million by the end of this century.

Average global warming is almost certain to exceed 1°C by the 2050s and may be as much as 6°C by the end of this century. Warming in the higher latitudes and polar regions is likely to be higher than the global averages.

Forecasts at the local level, including extreme events such as floods and cyclones, are less certain. Modelling the extraordinary complexity of the global climate system has to take into account shifts in regional systems, such as mid-latitude high pressures, and the local topography, such as the rain shadow in the lee of mountain ranges. However, climate scientists are beginning to project shifts in regional variables such as sunshine, humidity, precipitation, and wind.

Today's climate is the result of the cumulative emissions of greenhouse gases; today's emissions will have long-lasting effects on atmospheric concentrations, the Earth's energy balance and the climates of the future.

Predictions are compiled by running thousands of climate models, each run based on a slightly different set of assumptions. Each climate model is tested against the climate of the past 100 years as a way of judging its success in modelling future climate. Such modelling requires enormous computing power, but the results shown were derived by harnessing the spare capacity of tens of thousands of personal computers around the world. Almost all land areas are likely to get warmer. Globally, precipitation is likely to increase, but the pattern of where it will get wetter or drier is uncertain. For example, June to August is more likely to be wetter in East Asia and East Africa and drier in the Mediterranean (see below), while scenarios for central North America include both wetter and drier summers.

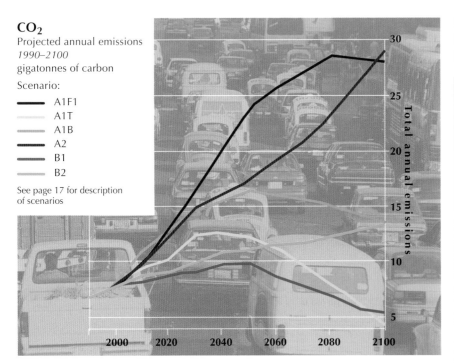

CO_2
Projected annual emissions
1990–2100
gigatonnes of carbon

Scenario:

— A1F1
— A1T
— A1B
— A2
— B1
— B2

See page 17 for description of scenarios

Total annual emissions

30
25
20
15
10
5

2000   2020   2040   2060   2080   2100

Scientific confidence in predictions of climate change *2007*

Virtually certain

Higher temperatures: warmer days, more frequent hot days and nights, fewer cold days and nights, over most land areas.

Very likely

More frequent warm spells and heat waves, over most land areas. More intense rain, over most areas.

Likely

Increased area affected by drought.
More frequent intense tropical cyclones: risk of stronger winds and heavier rain.
More frequent coastal flooding due to extreme high sea levels.

Forecasting Future Climates

Global temperatures are predicted
to continue rising.

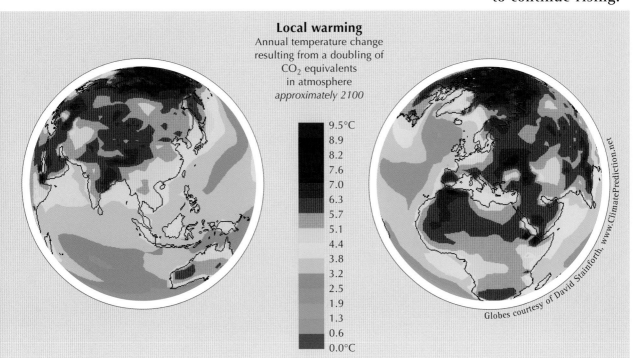

Local warming
Annual temperature change
resulting from a doubling of
CO_2 equivalents
in atmosphere
approximately 2100

9.5°C
8.9
8.2
7.6
7.0
6.3
5.7
5.1
4.4
3.8
3.2
2.5
1.9
1.3
0.6
0.0°C

Globes courtesy of David Stainforth, www.ClimatePrediction.net

Precipitation
Likelihood of wetter or drier June to August
with a doubling of CO_2 equivalents
in atmosphere
approximately 2100

━━━ Mediterranean Basin
━━━ Central North America
━━━ East Africa
━━━ East Asia

very likely

unlikely

drier ← → wetter
no
change

Projected warming
Increase in average
global temperatures
2000–2100
degrees Celsius

▨ average of global
climate models

▨ area of uncertainty

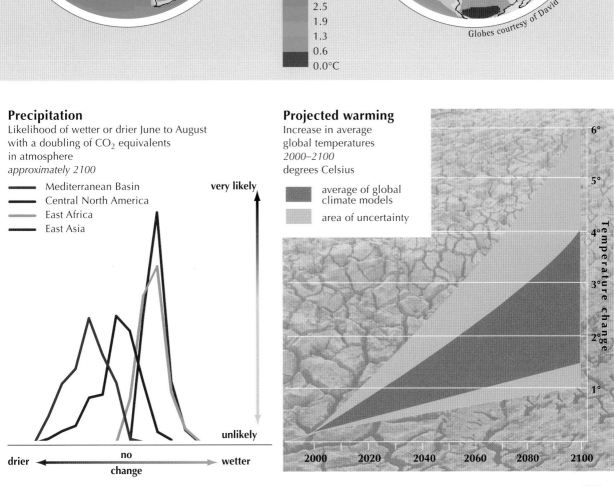

6°
5°
4°
3°
2°
1°

Temperature change

2000   2020   2040   2060   2080   2100

37

PART 3 DRIVING CLIMATE CHANGE

The climate system is relentlessly changing, and change may be rapid. The immediate causes of this change are the emissions of greenhouse gases released from energy production and consumption, agriculture, transport and ecological processes. Behind these sources of greenhouse gas emissions are broader driving forces related to economic transformations, prospects for alternative energy pathways and equity across regions and populations. For many, emissions of greenhouse gases are necessary for survival; for others, they result from luxury consumption and lifestyles.

The last few centuries have seen startling transformations in human history and, especially, how we use our environment. The Industrial Revolution led to vast improvements in livelihoods and a rapid growth in greenhouse gas emissions as fossil fuels, beginning with coal, were used more intensively. With the internal combustion engine and transportation, new technologies fed demand and welfare, and further accelerated emissions. Transportation of market goods, technological capabilities and growing populations contribute to changing land use and deforestation.

Historically, industrialized nations relying on fossil fuels have played a major role in increasing concentrations of carbon dioxide, and they remain the key emitters today. Focusing on energy needs for economic development, rather than for sustainability, industrialized countries invested heavily in carbon-intensive technologies, such as coal-fired power plants, massive road systems, and electrical grids. Globalization has made extensive transportation of goods and services the norm, and access to foreign markets crucial for economic growth.

Our future is partly dependent on the path established by past developments. With some greenhouse gases remaining in the atmosphere for centuries, historical emissions have guaranteed the inevitability of climate change for several decades, regardless of policy responses. And the burden of impacts is likely to be most serious in developing countries, even though they have contributed little to historical emissions. This imbalance between responsibility for the current causes of climate change and its impacts creates an enduring global inequity. While the economies of developing countries such as China and India are growing rapidly, leading to greater energy use and higher living standards, other countries cannot afford national electricity grids, much less energy-intensive luxuries.

The strength and momentum of these driving forces, the entrenched commitment to carbon-intensive economies, and the relationship between luxury consumption and basic needs are all part of the challenges in achieving reductions in emissions.

"The times of simply burning oil, one of our most important resources, are coming to an end."

Dr Kajo Schommer, former Economics Minister of Saxony, Germany, 2001

CUMULATIVE CARBON EMISSIONS

Share of total emissions of carbon dioxide (CO_2) from fossil fuel burning and cement production
1950–2000

1% of total emissions:
9,633 million tonnes
CO_2

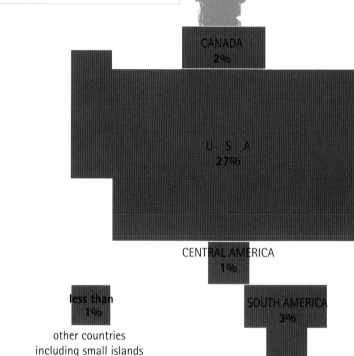

CANADA
2%

U S A
27%

CENTRAL AMERICA
1%

less than
1%
other countries
including small islands

SOUTH AMERICA
3%

Carbon dioxide accounts for half of global warming. It can remain in the atmosphere for up to 200 years. About half of all current greenhouse gas emissions are from the energy used in heating and lighting, transportation and manufacturing. Countries with a long history of industrialization have contributed the majority of the greenhouse gases in the atmosphere.

In the future, countries currently undergoing industrialization will account for a high percentage of annual greenhouse gas emissions, but it will be many years before their accumulated emissions match those of today's most industrialized nations.

This difference in past and future contributions to the overall levels of greenhouse gases raises important equity issues that are at the heart of international negotiations over how best to mitigate and adapt to climate change.

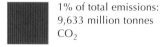

CO_2 in the atmosphere
Global concentration
1870–2005
parts per million

2005
381ppm

380	
360	
340	
320	
300	
280	

1870 1880 1890 1900 1910 1920 1930 1940 1950 1960 1970 1980 1990 2000 2010

Emissions Past and Present

Most greenhouse gases have been, and are, emitted to meet the needs of modern industrial societies.

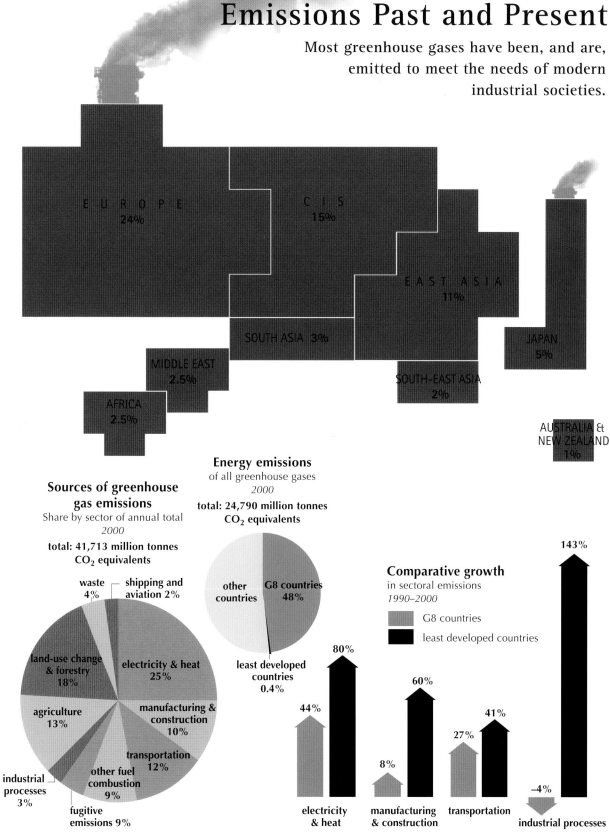

EUROPE
24%

CIS
15%

EAST ASIA
11%

SOUTH ASIA 3%

MIDDLE EAST
2.5%

AFRICA
2.5%

SOUTH-EAST ASIA
2%

JAPAN
5%

AUSTRALIA &
NEW ZEALAND
1%

Sources of greenhouse gas emissions
Share by sector of annual total
2000

total: 41,713 million tonnes CO$_2$ equivalents

waste 4%
shipping and aviation 2%
land-use change & forestry 18%
electricity & heat 25%
agriculture 13%
manufacturing & construction 10%
transportation 12%
industrial processes 3%
other fuel combustion 9%
fugitive emissions 9%

Energy emissions
of all greenhouse gases
2000

total: 24,790 million tonnes CO$_2$ equivalents

other countries
G8 countries 48%
least developed countries 0.4%

Comparative growth
in sectoral emissions
1990–2000

G8 countries
least developed countries

143%

80%

60%

44%

41%

27%

8%

–4%

electricity & heat

manufacturing & construction

transportation

industrial processes

FOSSIL FUEL BURNING

Annual carbon dioxide (CO_2) emissions from the burning of oil, natural gas and coal
2002
million tonnes

- 1,000 and above
- 500 – 999
- 100 – 499
- 50 – 99
- below 50
- no data

Change in annual CO_2 emissions
2002 compared with 1975

↑ increased by 300% or more

↓ decreased

Nearly two-thirds of carbon dioxide emissions, along with a significant amount of nitrous oxide and methane, derive from the burning of fossil fuels such as oil, natural gas, and coal. These are burned primarily for electricity, transportation, heating and cooling, and industrial processes.

Many industrialized nations are beginning to curb their carbon dioxide emissions, by more efficient fuel use, by using alternative sources of fuel, or simply because of economic slowdown. In many of the newly industrializing countries, however, emissions have increased markedly over recent decades, although their emissions per person are still relatively very low.

While petroleum reserves are limited, there are still hundreds of years of coal reserves worldwide, which could remain a significant source of greenhouse gas emissions. China, for example, depends on coal for over 75 percent of its total energy, and the decisions it takes over future power generation will have a major impact on levels of atmospheric carbon dioxide.

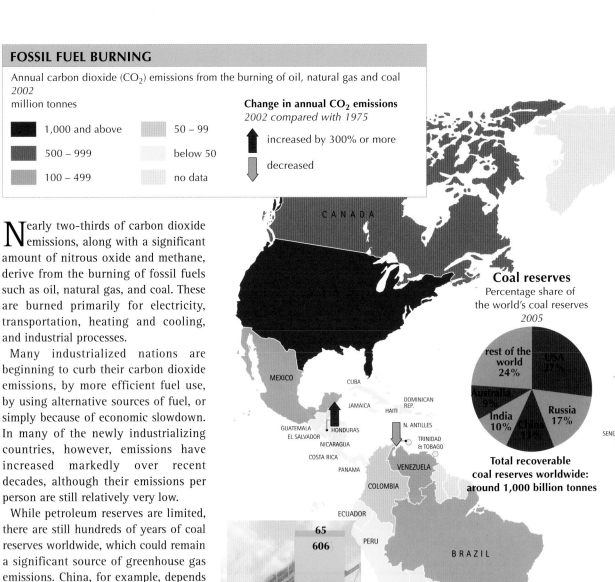

Coal reserves
Percentage share of
the world's coal reserves
2005

- rest of the world 24%
- USA 27%
- Australia 9%
- India 10%
- China 13%
- Russia 17%

**Total recoverable
coal reserves worldwide:
around 1,000 billion tonnes**

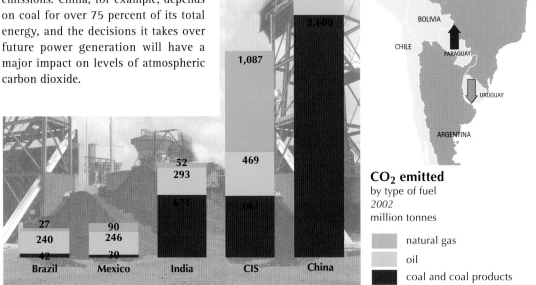

CO_2 emitted
by type of fuel
2002
million tonnes

- natural gas
- oil
- coal and coal products

	Brazil	Mexico	India	CIS	China
natural gas	27	90	52	469	65
oil	240	246	293	661	606
coal and coal products	42	30	671	1,087	2,600

Fossil Fuels

The emission of greenhouse gases from the burning of fossil fuels is the major cause of climate change.

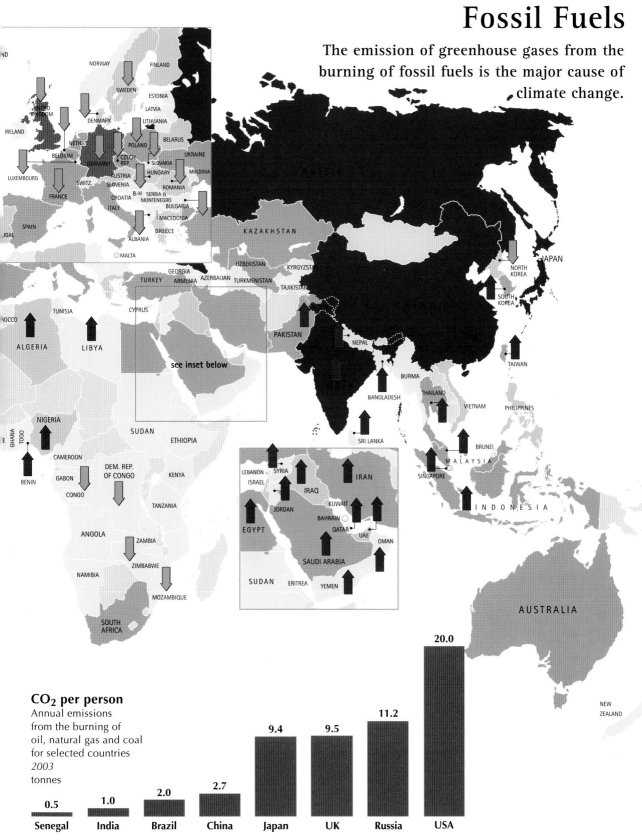

CO₂ per person
Annual emissions from the burning of oil, natural gas and coal for selected countries
2003
tonnes

Senegal	India	Brazil	China	Japan	UK	Russia	USA
0.5	1.0	2.0	2.7	9.4	9.5	11.2	20.0

METHANE EMISSIONS

Annual emissions per person
2000
tonnes of CO_2 equivalents

■ 5.0 and above

■ 2.5 – 4.9   ▨ below 1.0

■ 1.0 – 2.4   ▨ no data

Methane hydrates

A vast amount of methane (an estimated 20 million trillion m^3 worldwide) is trapped in permafrost ice and under-sea sediments in a form known as methane hydrates or clathrates. Its release into the atmosphere would be catastrophic, but there is uncertainty about what would trigger a mass release and how much of the methane would be transformed into CO_2 by sea water before it reached the atmosphere.

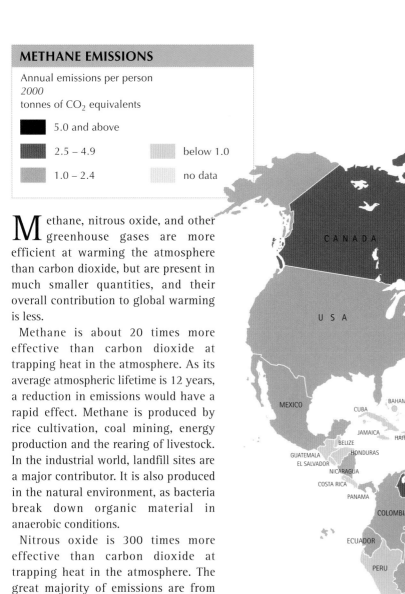

M ethane, nitrous oxide, and other greenhouse gases are more efficient at warming the atmosphere than carbon dioxide, but are present in much smaller quantities, and their overall contribution to global warming is less.

Methane is about 20 times more effective than carbon dioxide at trapping heat in the atmosphere. As its average atmospheric lifetime is 12 years, a reduction in emissions would have a rapid effect. Methane is produced by rice cultivation, coal mining, energy production and the rearing of livestock. In the industrial world, landfill sites are a major contributor. It is also produced in the natural environment, as bacteria break down organic material in anaerobic conditions.

Nitrous oxide is 300 times more effective than carbon dioxide at trapping heat in the atmosphere. The great majority of emissions are from agriculture – nitrogen-based fertilizers and livestock manure – with additional releases in waste, industrial processes and energy use.

Manufactured gases such as halocarbons – including chlorofluorocarbons (CFCs), hydrofluorocarbons (HFCs), and perfluorocarbons (PFCs) – and compounds such as sulfur hexafluoride (SF_6) have long lifetimes in the atmosphere. Sulfur hexafluoride is used as an insulator for circuit breakers, and to stop oxidation of molten magnesium during processing. HFCs are used in refrigeration units, in place of CFCs.

Nitrous oxide
Regional share
of total emissions
1990

Oceania 6%
South America 15%
Asia 31%
Africa 14%
North America 18%
Europe 17%

Global warming potential
The amount of heat trapped by gas over a standard period of time. The GWP for CO_2 is the reference unit.

carbon dioxide 1
methane 23
nitrous oxide 296

120

Methane and Other Gases

A range of greenhouse gases contribute to climate change.

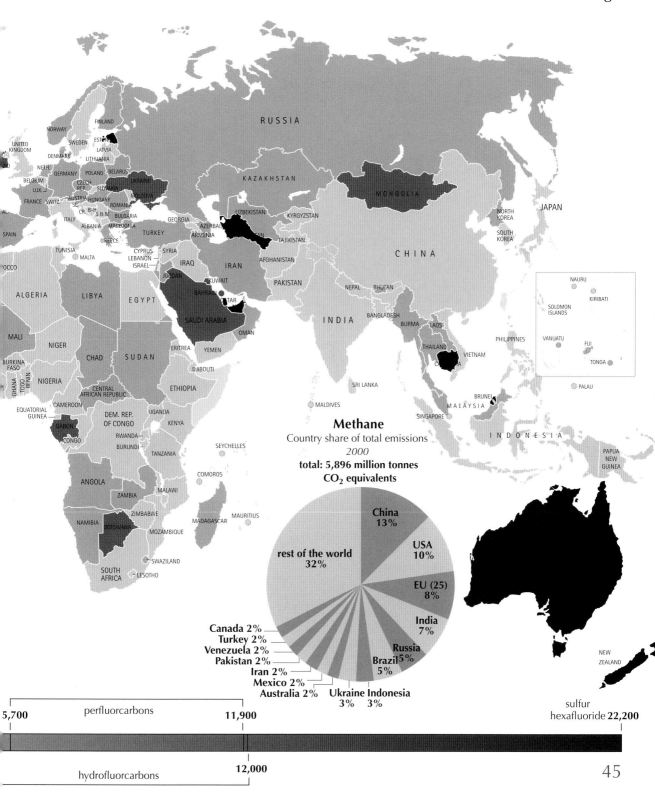

Methane
Country share of total emissions
2000
total: 5,896 million tonnes CO$_2$ equivalents

China 13%
USA 10%
EU (25) 8%
India 7%
Russia 5%
Brazil 5%
rest of the world 32%
Canada 2%
Turkey 2%
Venezuela 2%
Pakistan 2%
Iran 2%
Mexico 2%
Australia 2%
Ukraine 3%
Indonesia 3%

5,700   perfluorocarbons   11,900

sulfur hexafluoride 22,200

hydrofluorcarbons   12,000

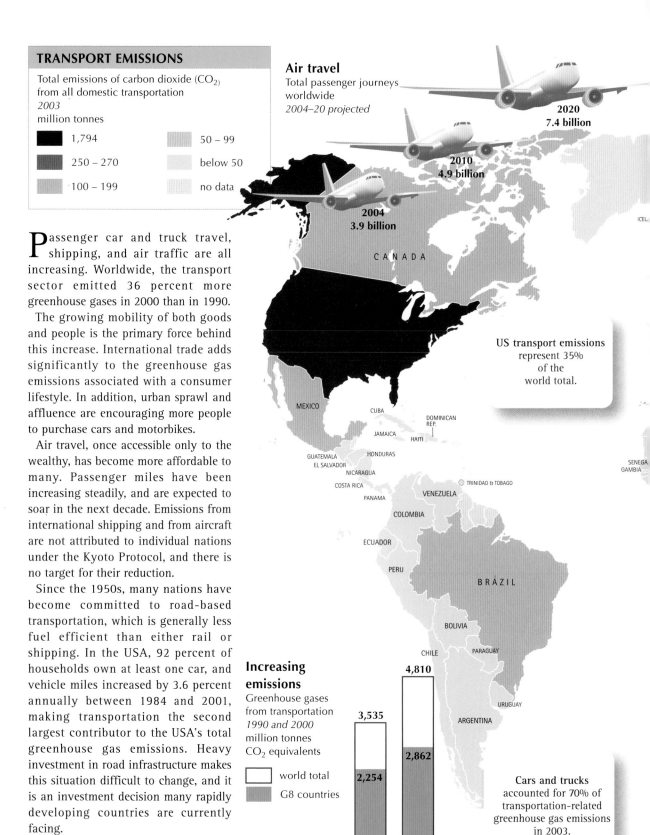

TRANSPORT EMISSIONS

Total emissions of carbon dioxide (CO_2)
from all domestic transportation
2003
million tonnes

1,794	50 – 99
250 – 270	below 50
100 – 199	no data

Air travel
Total passenger journeys
worldwide
2004–20 projected

2020
7.4 billion

2010
4.9 billion

2004
3.9 billion

CANADA

US transport emissions
represent 35%
of the
world total.

MEXICO
CUBA
DOMINICAN REP.
JAMAICA
HAITI
GUATEMALA
EL SALVADOR
HONDURAS
NICARAGUA
COSTA RICA
PANAMA
TRINIDAD & TOBAGO
VENEZUELA
COLOMBIA
ECUADOR
PERU
BRAZIL
BOLIVIA
CHILE
PARAGUAY
URUGUAY
ARGENTINA
ICEL.
SENEGAL
GAMBIA

Passenger car and truck travel,
shipping, and air traffic are all
increasing. Worldwide, the transport
sector emitted 36 percent more
greenhouse gases in 2000 than in 1990.

The growing mobility of both goods
and people is the primary force behind
this increase. International trade adds
significantly to the greenhouse gas
emissions associated with a consumer
lifestyle. In addition, urban sprawl and
affluence are encouraging more people
to purchase cars and motorbikes.

Air travel, once accessible only to the
wealthy, has become more affordable to
many. Passenger miles have been
increasing steadily, and are expected to
soar in the next decade. Emissions from
international shipping and from aircraft
are not attributed to individual nations
under the Kyoto Protocol, and there is
no target for their reduction.

Since the 1950s, many nations have
become committed to road-based
transportation, which is generally less
fuel efficient than either rail or
shipping. In the USA, 92 percent of
households own at least one car, and
vehicle miles increased by 3.6 percent
annually between 1984 and 2001,
making transportation the second
largest contributor to the USA's total
greenhouse gas emissions. Heavy
investment in road infrastructure makes
this situation difficult to change, and it
is an investment decision many rapidly
developing countries are currently
facing.

Increasing emissions
Greenhouse gases
from transportation
1990 and 2000
million tonnes
CO_2 equivalents

world total	
G8 countries	

3,535

4,810

2,254

2,862

1990   2000

Cars and trucks
accounted for 70% of
transportation-related
greenhouse gas emissions
in 2003.

Transportation

International trade, travel and a growing dependence on motor vehicles make transportation one of the main sources of greenhouse gas emissions.

International shipping and aviation
Emissions
1975–2003
million tonnes CO_2

shipping
aviation

shipping: 326 (1975), 291 (1985), 364 (1990), 405 (1995), 459 (2003)

aviation: 176 (1975), 230 (1985), 286 (1990), 296 (1995), 359 (2003)

1975   1985   1990   1995   2003

The carbon cycle is an essential part of life on earth. Carbon is stored in the form of plants, in the soil and dissolved in the oceans. The absorption of carbon from the atmosphere by plants and soil organisms, and its release through waste and decay, is part of the natural carbon cycle. When people cut down forests to make way for intensive cultivation, or simply to build on, more carbon is released than is absorbed. The natural carbon balance is disrupted.

During the 20th century and, in particular, since the 1950s, deforestation in South America, Africa and parts of Asia has released large amounts of carbon. Farming practices in dry regions, such as in west and east Africa, and India, have also led to the release of carbon from the soil. Over the last century in North America and Europe there was much less change in land use, and forests and tree plantations serve as carbon sinks on a modest scale. About a quarter of the carbon released into the atmosphere over the last 150 years has arisen from a change in the way land is used.

Oceans also exchange carbon dioxide with the atmosphere, and store a vast amount of carbon in their deeper layers. Dead plants and animals containing carbon sink to the ocean floor, thus removing carbon from the atmosphere. If oceans become significantly warmer, they may start to release more carbon than they absorb. A consequence of increased concentrations of atmospheric carbon dioxide, irrespective of climate change, is an increase in ocean acidity. Ocean chemistry is changing a hundred times more rapidly than it has done in the past 100,000 years, and acidity may reach levels that have not occurred for tens of millions of years. The dissolved carbon dioxide, carbonic acid, is corrosive to calcium carbonate shells and exoskeletons. Even low projections for future carbon dioxide emissions indicate that corals could be rare on tropical and subtropical reefs by 2050.

Carbon release

Net carbon released worldwide from terrestrial ecosystems as a result of changes in land use
average across 5 years: 1896–1900; 1945–50; 1996–2000
million tonnes

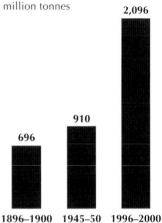

1896–1900	1945–50	1996–2000
696	910	2,096

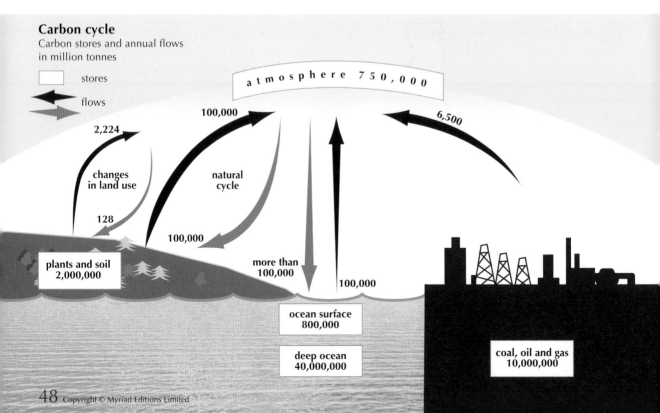

Carbon cycle

Carbon stores and annual flows in million tonnes

☐ stores
◀ flows

atmosphere 750,000

2,224
changes in land use
128

100,000
natural cycle

6,500

100,000

more than 100,000

100,000

plants and soil 2,000,000

ocean surface 800,000

deep ocean 40,000,000

coal, oil and gas 10,000,000

Disrupting the Carbon Balance

Carbon is essential in the natural environment,
but changes in land use may release stored
carbon and contribute to climate change.

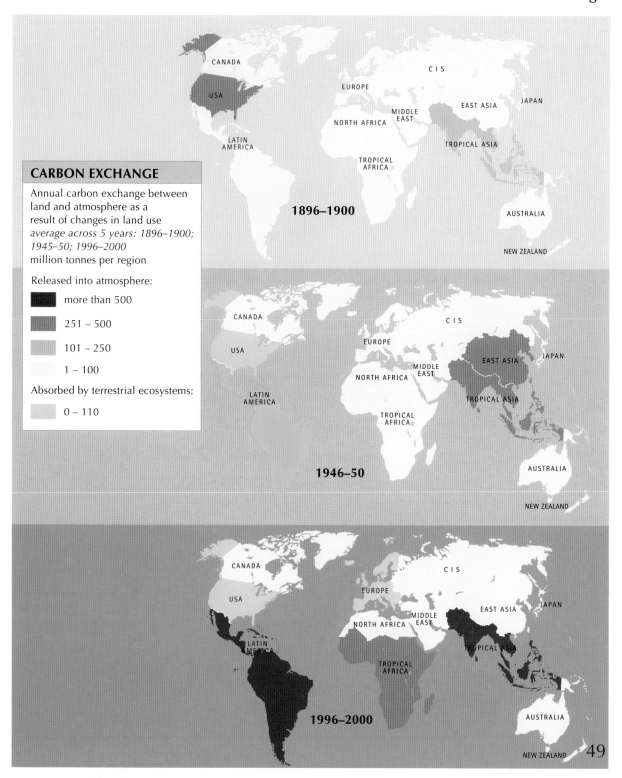

CARBON EXCHANGE

Annual carbon exchange between
land and atmosphere as a
result of changes in land use
*average across 5 years: 1896–1900;
1945–50; 1996–2000*
million tonnes per region

Released into atmosphere:

more than 500

251 – 500

101 – 250

1 – 100

Absorbed by terrestrial ecosystems:

0 – 110

1896–1900

1946–50

1996–2000

49

AGRICULTURE

Greenhouse gas emissions from agriculture
2000
million tonnes carbon dioxide (CO_2) equivalents

- 440 and above
- 100 – 149
- 50 – 99
- 25 – 49
- below 25
- no data

agriculture represents 25% or more of GDP and produces emissions over 10 million tonnes

Agriculture accounts for about a third of global emissions of the greenhouse gases carbon dioxide (CO_2), methane, and nitrous oxide, but in many developing countries it is the main economic activity of the rural population. It is essential to meet basic needs: food, employment and income.

In wealthier countries, agriculture is primarily a commercial activity, accounting for a smaller proportion of the economy. Agriculture responds to consumer demands for a year-round choice of fresh fruit and vegetables.

Plans for reducing emissions in agriculture must consider the consequences for less advantaged populations. The growing of rice in flooded fields releases methane from waterlogged soils, but rice feeds a third of the world's population and is the staple diet of many poor people in Asia. Livestock are also a source of methane, but those on the small holdings of poor farmers and pastoralists produce much less gas than the well-fed cattle in large-scale commercial enterprises. Fertilizers produce nitrous oxide, but also provide a much-needed boost to food production in some areas.

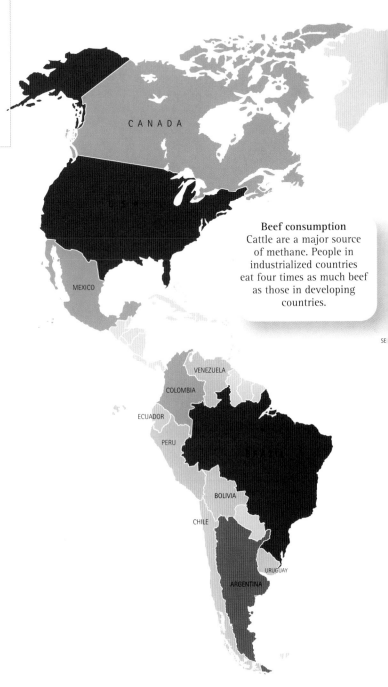

CANADA

Beef consumption
Cattle are a major source of methane. People in industrialized countries eat four times as much beef as those in developing countries.

MEXICO

VENEZUELA
COLOMBIA
ECUADOR
PERU
BOLIVIA
CHILE
URUGUAY
ARGENTINA

SE

Agriculture

Greenhouse gases are emitted in the production of food. While some agriculture meets basic needs, some simply provides wealthy consumers with the luxury of choice.

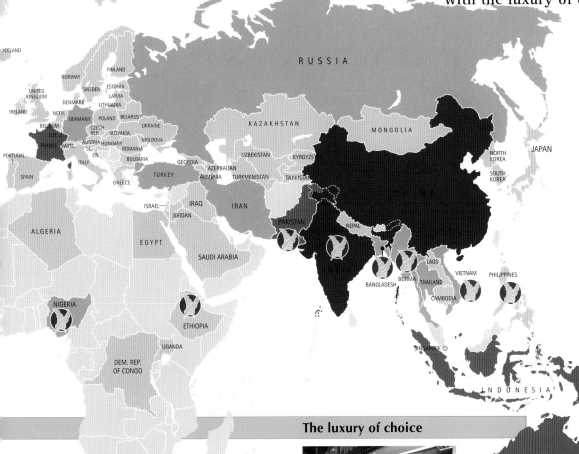

ICELAND

NORWAY
FINLAND
UNITED KINGDOM
SWEDEN
ESTONIA
DENMARK
LATVIA
LITHUANIA
IRELAND
NETH.
GERMANY
POLAND
BELARUS
BELGIUM
LUX.
CZECH REP.
SLOVAKIA
UKRAINE
FRANCE
SWITZ.
AUSTRIA
HUNGARY
MOLDOVA
SL.
CR.
ROMANIA
PORTUGAL
ITALY
BULGARIA
SPAIN
GREECE
TURKEY
GEORGIA
AZERBAIJAN
ARMENIA
TURKMENISTAN
TAJIKISTAN
ISRAEL
IRAQ
JORDAN
IRAN
ALGERIA
EGYPT
SAUDI ARABIA

RUSSIA

KAZAKHSTAN
UZBEKISTAN
KYRGYZ.
MONGOLIA
NORTH KOREA
SOUTH KOREA
JAPAN
CHINA
PAKISTAN
NEPAL
INDIA
BANGLADESH
BURMA
LAOS
THAILAND
VIETNAM
PHILIPPINES
CAMBODIA
NIGERIA
ETHIOPIA
UGANDA
DEM. REP. OF CONGO
SINGAPORE
INDONESIA
SOUTH AFRICA
AUSTRALIA
NEW ZEALAND

The luxury of choice

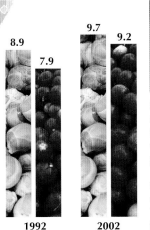

CO$_2$ emissions
from transporting food consumed in the UK
1992 and 2002
million tonnes

8.9
7.9
9.7
9.2

within the UK

from overseas

1992   2002

Consumers in industrialized countries expect an ever wider choice of food products throughout the year. The environmental costs associated with this are immense, especially in terms of greenhouse gas emissions.

In the UK, the emissions created by transporting food into and around the country, and by people driving to the shops, increased by 13% from 1992 to 2002, with imports seeing the largest growth.

PART 4   EXPECTED CONSEQUENCES

There is a growing confidence in the international scientific community that we are witnessing the first serious impacts of climate change. Warming trends are affecting ecosystems and resources in ways that are unprecedented. Some disasters, such as heat waves, have been attributed to the increased concentrations of greenhouse gases in the atmosphere.

The consequences of climate change occur everywhere, and are globally interconnected. The impacts of climate change on agriculture may begin with changes in crop yields at a local scale, but are ultimately felt around the world in higher prices and transportation costs. Vast subsidies in many industrialized countries ensure farmers in these countries remain in business; poor farmers in developing countries with little support are highly vulnerable.

Even something that seems relatively simple at a regional level, like switching to a new crop on the wealthy Great Plains of the USA, quickly becomes more complex on a global scale. Switching to a crop better suited to new conditions might require huge investment in planting and harvesting equipment, new storage and distribution infrastructure, and different farm labor arrangements. Elsewhere in the world, farmers who had been producing that crop will be faced with a major new competitor as well as the effects of climatic changes. Food manufacturers will find alternative suppliers, and consumers will face changes in price, quality, and availability. Federal governments will be called on to reduce the impacts of all such changes.

Sea-level rise and reduced sea-ice coverage in the Arctic are other examples of how one thing may lead to another. Ports compete with each other to draw in economic development and employment. If warming in the Arctic opens a new shipping route, competitive advantages among ports will change. In the meantime, all port facilities will have to adjust to a rising sea level. In developing countries, where ports are key to international trade, foreign currency, and economic development, making those improvements will be difficult to afford.

The dire warnings of the collapse of entire societies cannot be discounted, but neither can we provide reliable forecasts of catastrophe. We do know that the burden of impacts falls most heavily on the poorer and more marginalized people. Nearly a billion people today live in poverty, without access to clean water and lacking basic health care, many of them dependent on agriculture. Diverting national funds to maintain key economic resources, like ports, will compete with the urgent demands for health care, education and other basic needs. Climate change is not solely an environmental issue; it has implications for achieving economic growth, human security, and broader social goals.

"We know that dramatic changes are still to come and they will result in huge economic costs. When we push an alarm button at IUCN, it's because there is a real cause for alarm."

Brett Orlando
IUCN Climate
Change Advisor,
2003

BIODIVERSITY AT RISK

Ecosystems and species
threatened by climate change
2005

Climate change will accelerate the extinction of a wide range of species. Some of the regions richest in biodiversity are already being affected. Catastrophic events, such as droughts, or even small changes in average temperature, can disrupt ecosystems built on the inter-dependency of thousands of species. Sea-level rise is destroying low-lying habitats.

Unprecedented rates of migration by both plants and animals will be needed if species are to keep up with climate change. While grasslands and desert can spread fairly quickly into new areas, slower-growing forests may find themselves outpaced in the race against time. In mountainous regions, some species may only have to move a few hundred meters up hill to find new terrain, but a problem arises when those already at the top of the mountain or on the most northerly landmasses in the Arctic need to move to cooler climes.

Some species have their escape routes blocked off, surrounded as they are by agricultural and urban development. The creation of migration corridors is a focus of much conservation work.

The annual destruction of an estimated 14.2 million hectares of tropical forest is of particular concern. The Earth's forests play a role in the natural process of storing carbon, while also providing habitat protection. Forest destruction accelerates both the impacts of climate change and the loss of biodiversity.

Between 20% and 30% of plant and animal species assessed so far are likely to be at increased risk of extinction if increases in global average temperature exceed 1.5°–2.5° C.

IPCC Working Group II,
2007

At least 40% of the world's economy and 80% of the needs of the poor are derived from biological resources.

Secretariat of the Convention
on Biological Diversity

The loss of each additional species reduces the options for nature and people to respond to changing conditions.

Jeffrey McNeely,
IUCN Chief Scientist

Non-native species

Climate change is expected to make ecosystems susceptible to invasion by non-native species such as the red fire ant, which threatens to destroy native flora and fauna in the southeastern USA.

Central America

The mountain forests of Central America are home to many endemic mammals, amphibians and birds as well as to 17,000 plant species. The region is also a vital corridor for many migrant bird species. Climate change, in particular a reduction in rainfall, may threaten this rich and unique habitat.

Brazilian Cerrado

This savannah (grassland) originally covered more than 20% of Brazil. Now it is so fragmented that it is unlikely that its 10,000 plant species will all be able to disperse to climatically suitable areas.

Global warming

Areas where ecosystems will change, for a scenario of a global mean temperature increase of 3°C.

The effect of warmer temperatures will lead to a general shift of ecosystems towards the poles. In some areas there is no available land at higher latitude, which will probably lead to the disappearance of an ecosystem.

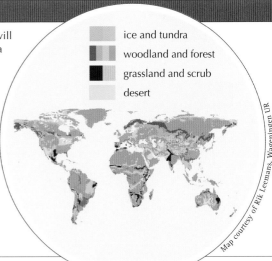

- ice and tundra
- woodland and forest
- grassland and scrub
- desert

Map courtesy of Rik Leemans, Wageningen UR

Disrupted Ecosystems

Many species and ecosystems, already at risk from human development, may not be able to adapt to new climatic conditions and stresses.

Europe

The ability of European birds and plants to migrate in the face of warmer temperatures is severely limited by the density of the human population, and by intensive cultivation. The Spanish Imperial Eagle, for example, currently largely restricted to nature parks and reserves, may find that unspoilt areas with suitable temperatures are just not available. The Scottish Crossbill, endemic to Scotland, UK, might be faced with a move to Iceland – a journey it would be unlikely to undertake successfully.

Mountains of Central Asia

Global warming poses a long-term threat to the dry terrain of the Tien Shan and Pamir mountains, characterized by high plateaus or steep-sided gorges, neither of which make plant migration easy. The destabilizing effects of the melting of frozen upper slopes may lead to the decline or disappearance of many endemic species.

Sundarbans Delta

This huge area of mangroves in the Bay of Bengal, home to the **Bengal tiger** and other important species, is threatened by rising sea levels.

Ocean warming

By allowing the spread of disease agents and parasites, ocean warming may contribute to the outbreak of species-based epidemics, such as the one that killed thousands of **striped dolphins** in the Mediterranean in the early 1990s.

Polynesia and Micronesia

The movement of endemic species is highly restricted on these widely scattered island groups, which are threatened by a rising sea level.

Mounts Kenya and Kilimanjaro

Warmer temperatures are causing the glaciers on these mountains to melt. This threatens the survival of endemic alpine species on the mountain slopes, and of streamside habitats that survive on the surrounding desert floor, nourished by the glacial melt water.

South Africa

In the Western Cape, a family of endemic flowering plants that includes the **golden pagoda**, only discovered in 1987, is facing possible extinction, as warmer temperatures severely restrict their habitat range. The Kruger National Park may lose up to 60% of the species currently protected.

Water is a vital resource and is often taken for granted. With populations increasing in some regions, and a rising demand for water to irrigate crops throughout the world, water supplies are already a cause for concern in many countries. It is now apparent that climate change might make the situation worse.

Some areas will experience less annual rainfall; in others it will be less predictable, with seasonal rains failing to materialize, or arriving with such ferocity that they create dangerous floods. An increase in temperature makes surface water evaporate more quickly, reducing supply and increasing demand, especially for water to irrigate crops. Warmer and longer summers also cause snow packs and glaciers to melt more quickly. More rapid melting increases river flows in the spring, but may reduce summer flows. Over the long term, a reduction in snow and ice may seriously threaten many river basins. For example, in northern India 500 million people rely on the Indus and Ganges, which are fed largely by glacial melt waters.

Rapid changes in weather patterns – from seasonal crises to a decade of low flows – leave people little time to adapt. Some regions will be forced to import more food, or even to import water itself; this has been suggested as an option for Israel and Southeast England. Water-intensive industries, such as paper and electronics manufacturing, will be unable to function, and economies will suffer as a consequence.

If water supplies fail completely, contaminated water, lack of hygiene and thirst will take their toll. Many of the effects of climate change can be countered by prioritizing the most urgent uses, adopting water-saving technology and more efficient irrigation methods. However, less-developed countries in drier parts of the world, which lack the technology and infrastructure to effectively manage their water resources, will suffer most.

Water use

Annual withdrawals per person and sectoral use
2000
cubic meters

- domestic
- agricultural
- industrial

There is a huge discrepancy in the amount of water used in different countries. Countries or regions reliant on heavily irrigated agriculture, or on industries requiring large quantities of water, may have to make adjustments as the changing climate reduces the amount of water available.

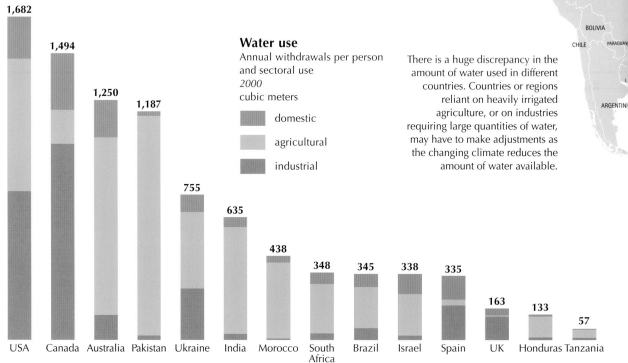

USA	Canada	Australia	Pakistan	Ukraine	India	Morocco	South Africa	Brazil	Israel	Spain	UK	Honduras	Tanzania
1,682	1,494	1,250	1,187	755	635	438	348	345	338	335	163	133	57

Threatened Water Supplies

Water scarcity is already a growing concern.
In some places climate change will make it
even more critical.

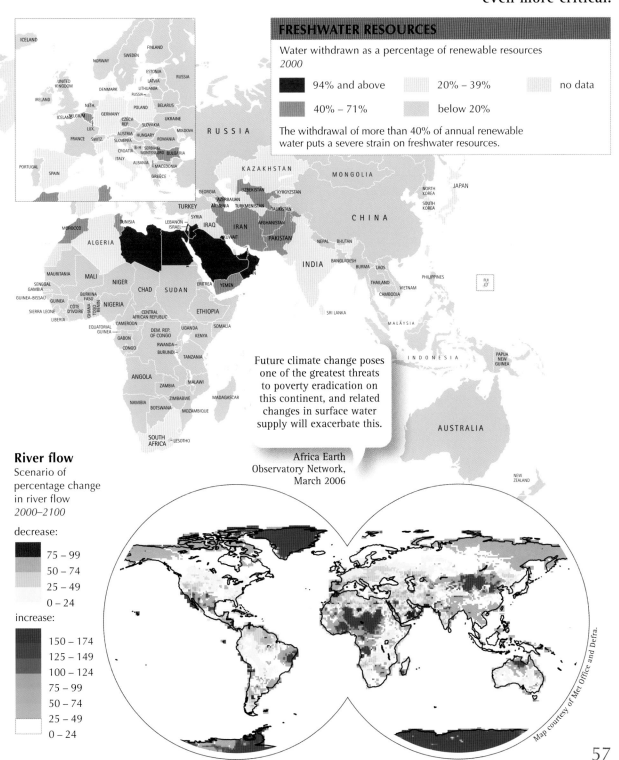

FRESHWATER RESOURCES

Water withdrawn as a percentage of renewable resources
2000

- 94% and above
- 40% – 71%
- 20% – 39%
- below 20%
- no data

The withdrawal of more than 40% of annual renewable
water puts a severe strain on freshwater resources.

Future climate change poses
one of the greatest threats
to poverty eradication on
this continent, and related
changes in surface water
supply will exacerbate this.

Africa Earth
Observatory Network,
March 2006

River flow
Scenario of
percentage change
in river flow
2000–2100

decrease:
- 75 – 99
- 50 – 74
- 25 – 49
- 0 – 24

increase:
- 150 – 174
- 125 – 149
- 100 – 124
- 75 – 99
- 50 – 74
- 25 – 49
- 0 – 24

Map courtesy of Met Office and Defra.

FUTURE FOOD PRODUCTION

Scenarios of regional change in cereal production
by 2050s–80s

 decrease (–8% to –19%)

small decrease (1% to –10%)

decrease or increase (–4% to 8%)

small increase (2% to 4%)

increase (1% to 20%)

Current malnutrition

25% or more of children
under 5 years moderately
or severely underweight
2005 or latest available data

The effect of climate change on agriculture depends on a combination of factors. Higher temperatures can stress plants, but also prolong growing seasons and allow a greater choice of crops to be grown. Higher concentrations of carbon dioxide speed growth and increase resilience to water stress. However, pests and diseases may also increase in response to the more benign climatic conditions. Demand for different foods, and changes in markets and trade regimes are also important factors.

Agriculture is highly adaptable. Crop calendars can be adjusted to avoid extreme hot periods, new varieties of plants can tolerate a range of conditions, and good soil management can overcome water stress. With economic incentives, world food production should not be adversely affected by climate change over the next 50 years or so. Agriculture in temperate climates may actually benefit from longer growing seasons and warmer temperatures.

In parts of the tropical and sub-tropical regions, however, reductions in rainfall, and increasing risk of drought, or of more intense rainfall and soil erosion, will severely affect agriculture. The capacity of developing countries to sustain agricultural production and food security is already challenged. The poor, those most likely to experience malnutrition, are likely to suffer further.

Food Security

Climate change threatens food security, although crop yields in temperate regions may improve.

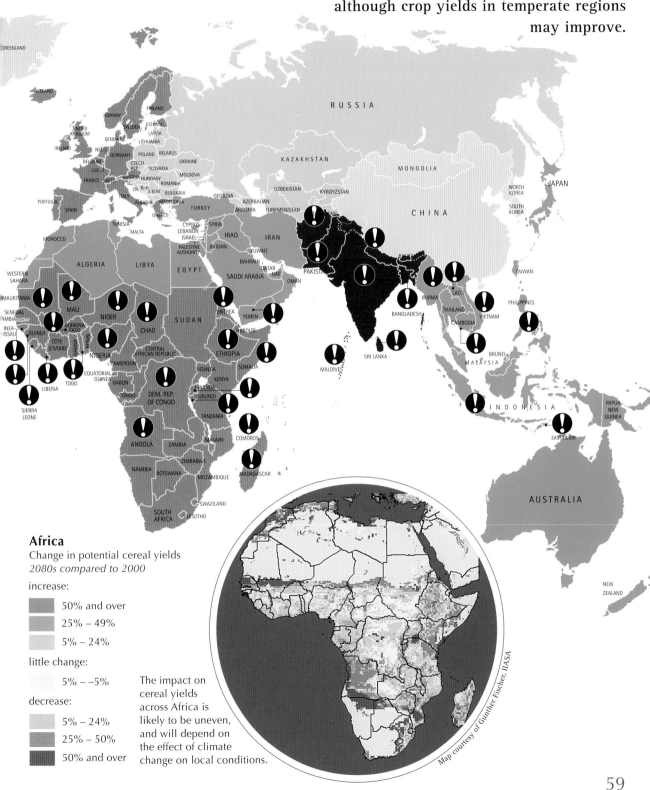

GREENLAND

ICELAND

NORWAY

FINLAND

SWEDEN

ESTONIA

LATVIA

LITHUANIA

UNITED KINGDOM

IRELAND

DENMARK

NETH.

BELGIUM

LUX.

GERMANY

POLAND

BELARUS

CZECH REP.

SLOVAKIA

UKRAINE

FRANCE

SWITZ.

AUSTRIA

HUNGARY

MOLDOVA

SL.

CR.

B-H

ROMANIA

ITALY

S.M.

BULGARIA

PORTUGAL

SPAIN

ALBANIA

MACEDONIA

GREECE

GEORGIA

RUSSIA

KAZAKHSTAN

MONGOLIA

NORTH KOREA

JAPAN

SOUTH KOREA

CHINA

TUNISIA

MALTA

TURKEY

CYPRUS

LEBANON

ISRAEL

SYRIA

AZERBAIJAN

ARMENIA

TURKMENISTAN

UZBEKISTAN

KYRGYZSTAN

MOROCCO

WESTERN SAHARA

ALGERIA

LIBYA

EGYPT

PALESTINE AUTHORITY

JORDAN

IRAQ

KUWAIT

IRAN

TAIWAN

MAURITANIA

SENEGAL

GAMBIA

GUINEA-BISSAU

MALI

BURKINA FASO

NIGER

CHAD

SUDAN

ERITREA

YEMEN

DJIBOUTI

SAUDI ARABIA

BAHRAIN

QATAR

UAE

OMAN

PAKISTAN

INDIA

BHUTAN

BANGLADESH

BURMA

LAOS

THAILAND

VIETNAM

PHILIPPINES

GUINEA

CÔTE D'IVOIRE

GHANA

BENIN

NIGERIA

CAMEROON

CENTRAL AFRICAN REPUBLIC

ETHIOPIA

SOMALIA

SRI LANKA

MALDIVES

CAMBODIA

BRUNEI

MALAYSIA

SIERRA LEONE

LIBERIA

TOGO

EQUATORIAL GUINEA

GABON

CONGO

UGANDA

KENYA

DEM. REP. OF CONGO

RWANDA

BURUNDI

TANZANIA

ANGOLA

ZAMBIA

MALAWI

COMOROS

INDONESIA

PAPUA NEW GUINEA

EAST TIMOR

NAMIBIA

ZIMBABWE

BOTSWANA

MOZAMBIQUE

MADAGASCAR

SWAZILAND

SOUTH AFRICA

LESOTHO

AUSTRALIA

NEW ZEALAND

Africa

Change in potential cereal yields
2080s compared to 2000

increase:

- 50% and over
- 25% – 49%
- 5% – 24%

little change:

- 5% – –5%

decrease:

- 5% – 24%
- 25% – 50%
- 50% and over

The impact on cereal yields across Africa is likely to be uneven, and will depend on the effect of climate change on local conditions.

Map courtesy of Gunther Fischer, IIASA

HEALTH IMPACT OF CLIMATE CHANGE

Annual number of DALYs per million people from malnutrition, diarrhea, flooding, and malaria caused by climate-related conditions
2000 by WHO region

- more than 3,000
- 1,500 – 3,000
- 100 – 199
- under 10
- no data

DALYs or Disability Adjusted Life Years are the number of years of potential life lost due to premature mortality, plus years of productive life lost due to disability.

Tick-borne diseases

Lyme disease, which can cause long-term disability, is spreading in the USA and Europe as winter temperatures warm and daily temperatures rise. The habitat favorable to the **deer ticks** that carry the disease is likely to spread. Rocky Mountain spotted fever, Q fever, and, especially in Europe, tick-borne encephalitis are among other diseases that may also spread as the climate changes.

Higher global temperatures are altering local climates: some areas may experience more variable rainfall patterns, warmer winters and drier summers. Where people are already vulnerable to disease as a result of poverty and malnutrition, even small changes in climate may have an effect on health.

Rainfall, temperature, and humidity have a major influence on the distribution of disease pathogens and pests. Warmer temperatures, longer growing seasons, the absence of pest-killing sub-zero temperatures, and increased rainfall all extend the habitat ranges for diseases, and for the insects, rodents and other organisms that carry them. Climate changes will favor the spread of diseases into previously unaffected areas.

While fewer people may die from cold, warmer weather may lead to increased heat stress. It may also lead to higher levels of air pollutants from forest fires in rural areas, and from the formation of ozone and volatile organic compounds in urban areas. The number of deaths related to respiratory conditions will rise.

Flooding increases the risk of waterborne diseases such as cholera, typhoid, and dysentery, and of mosquito-borne diseases, including malaria and yellow fever. Water scarcity and drought reduce food production, as well as contributing to the spread of diseases associated with poor water quality and lack of sanitation. The cumulative effects of environmental stresses, including malnutrition, further reduces the ability to fight off infections.

LATIN AMERICA
AND CARIBBEAN

Intestinal diseases

Heavy rain sometimes contaminates water supplies. Lack of water during a dry season or drought makes it difficult for people to keep clean, increasing the incidence of disease. Both extremes are likely to become more frequent.

Total health impact

Annual number of DALYs from malnutrition, diarrhea, flooding, and malaria caused by climate-related conditions
2000
by WHO region

The World Health Organization has estimated the effect of four global health threats – malnutrition, diarrhea, flooding, and malaria – in an assessment of the contribution of climate-related stress to the global disease burden. Further climate change is likely to make matters worse.

2,572,000 South-East Asia

Africa

768,000 Eastern Mediterranean

169,000 Western Pacific

92,000 Latin America & Caribbean

8,000 Developed countries

Threats to Health

Climate change threatens human health. The poorest regions are likely to be the hardest hit.

WESTERN PACIFIC

EASTERN MEDITERRANEAN

SOUTH-EASTERN ASIA

Malaria

Over 3.2 billion people are currently exposed to the risk of malaria, and over a million die from it each year. These numbers are likely to increase with climate change. Malaria is carried by **mosquitoes**, which, in tolerably warmer temperatures, mature more quickly, bite more often, reproduce more, and breed longer. Wetter conditions can also foster outbreaks of the disease. Climate change will allow malaria to spread into northern and upper mountain regions, if other ecological factors do not limit it.

Health effects of flooding

Flooding, a likely outcome of climate change in many parts of the world, brings a variety of health challenges. Water supplies may be contaminated and standing water creates breeding grounds for disease-carrying organisms such as mosquitoes. Soil transported by flood waters may move soil-borne diseases such as anthrax, and toxic contaminants such as heavy metals and organic chemicals, to previously unexposed areas. Mold developing on flooded property may contribute to respiratory problems.

Warmer temperatures cause oceans to expand, and melting glaciers add to the volume of water. Both result in rising sea levels. The mean sea level rose by around 15 centimeters during the 20th century, and projections indicate a further rise of between 20 and 90 centimeters between 1990 and 2100. Even if greenhouse gas emissions are radically reduced over the next decades, because of the huge thermal mass of the oceans the sea level will continue to rise for centuries – a long-term consequence of emissions already released.

A rise in mean sea level of 1 meter, at the upper range of estimates for the next hundred years, will have drastic consequences for many coastal communities. The Maldives islands in the Indian Ocean will be almost completely inundated, as will large parts of island groups in the Caribbean and Pacific. Around the world, valuable agricultural land will be lost, and cities will be threatened. With stronger windstorms possible, many low-lying communities will be at risk from storm surges, which can add 5 meters or more to the mean sea level. Coastal movement – sinking or rising – also affects the height of the sea relative to the land. The movement of seawater higher up rivers, and into freshwater aquifers, will affect drinking water supplies across the world, threatening the viability of many communities.

Far more serious sea-level rise is possible. Were the Greenland ice cap to melt, it would add an estimated 7 meters to the global sea level. The ice sheet covering West Antarctica rests on rock that is below sea level. Were it to melt, the sea level might rise by a further 5 meters. At present, there is only a low probability that these ice sheets will collapse in the next few centuries. However, if global warming exceeds 3°C or so, these scenarios become increasingly likely, with potentially catastrophic consequences around the world.

Total land loss
from sea-level rise
selected regions
by 2100
square kilometers

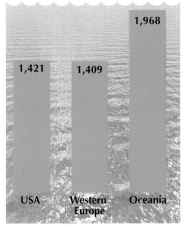

The figures for both chart and world map are based on a scenario of relatively high greenhouse gas emissions (A1F1), but assume some protection against sea-level rise and storm surges.

Projected sea-level rise
1990–2100

average of 35 projections of greenhouse gas emissions

area of uncertainty

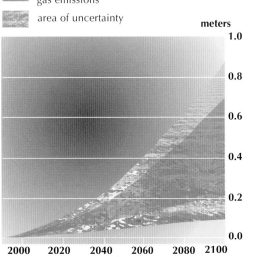

The estimates of sea-level rise vary because of differing projections of greenhouse gas emissions, models of the climate and oceans, and assumptions about land-ice melt.

Nile Delta

area inundated by a 1-meter sea-level rise

A 1-meter sea-level rise could affect 15% of Egypt's habitable land.

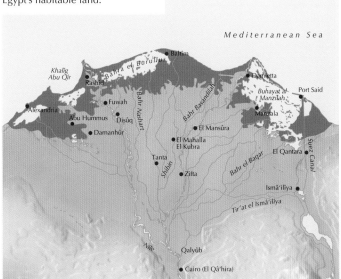

Rising Sea Levels

Thermal expansion of oceans and melting ice will lead to a substantial rise in sea level, threatening many coastal communities.

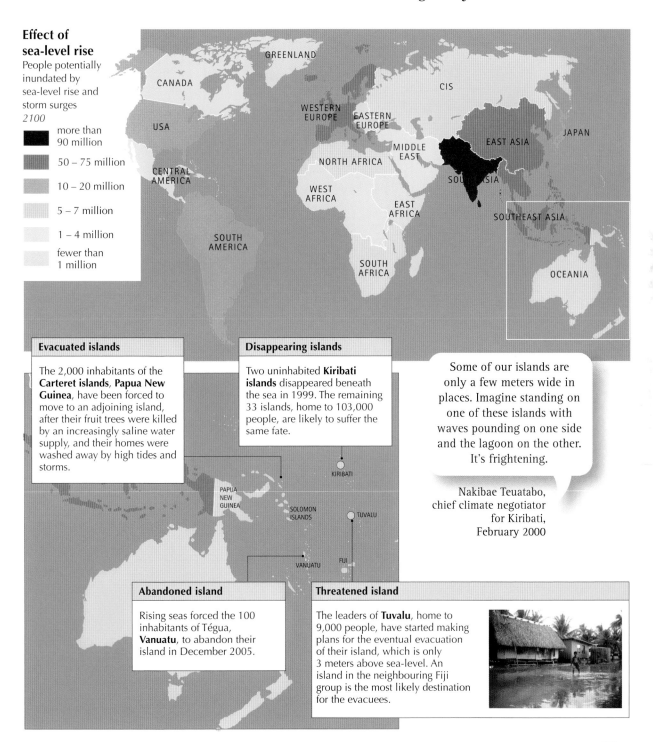

Effect of sea-level rise

People potentially inundated by sea-level rise and storm surges
2100

- more than 90 million
- 50 – 75 million
- 10 – 20 million
- 5 – 7 million
- 1 – 4 million
- fewer than 1 million

GREENLAND
CANADA
CIS
WESTERN EUROPE
EASTERN EUROPE
USA
MIDDLE EAST
EAST ASIA
JAPAN
NORTH AFRICA
CENTRAL AMERICA
SOUTH ASIA
WEST AFRICA
EAST AFRICA
SOUTHEAST ASIA
SOUTH AMERICA
SOUTH AFRICA
OCEANIA

PAPUA NEW GUINEA
KIRIBATI
SOLOMON ISLANDS
TUVALU
VANUATU
FIJI

Evacuated islands

The 2,000 inhabitants of the **Carteret islands**, **Papua New Guinea**, have been forced to move to an adjoining island, after their fruit trees were killed by an increasingly saline water supply, and their homes were washed away by high tides and storms.

Disappearing islands

Two uninhabited **Kiribati islands** disappeared beneath the sea in 1999. The remaining 33 islands, home to 103,000 people, are likely to suffer the same fate.

Some of our islands are only a few meters wide in places. Imagine standing on one of these islands with waves pounding on one side and the lagoon on the other. It's frightening.

Nakibae Teuatabo,
chief climate negotiator
for Kiribati,
February 2000

Abandoned island

Rising seas forced the 100 inhabitants of Tégua, **Vanuatu**, to abandon their island in December 2005.

Threatened island

The leaders of **Tuvalu**, home to 9,000 people, have started making plans for the eventual evacuation of their island, which is only 3 meters above sea-level. An island in the neighbouring Fiji group is the most likely destination for the evacuees.

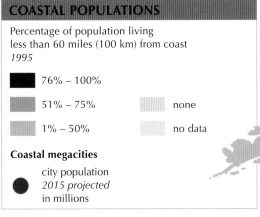

COASTAL POPULATIONS

Percentage of population living
less than 60 miles (100 km) from coast
1995

- 76% – 100%
- 51% – 75%
- 1% – 50%
- none
- no data

Coastal megacities

city population
2015 projected
in millions

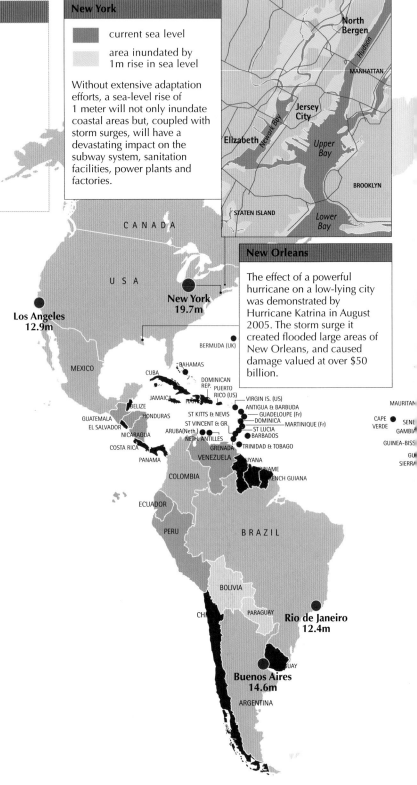

New York

- current sea level
- area inundated by 1m rise in sea level

Without extensive adaptation efforts, a sea-level rise of 1 meter will not only inundate coastal areas but, coupled with storm surges, will have a devastating impact on the subway system, sanitation facilities, power plants and factories.

North Bergen

Hudson

MANHATTAN

Jersey City

Elizabeth

Newark Bay

Upper Bay

BROOKLYN

STATEN ISLAND

Lower Bay

New Orleans

The effect of a powerful hurricane on a low-lying city was demonstrated by Hurricane Katrina in August 2005. The storm surge it created flooded large areas of New Orleans, and caused damage valued at over $50 billion.

CANADA

USA

New York
19.7m

Los Angeles
12.9m

BERMUDA (UK)

MEXICO

BAHAMAS

CUBA

DOMINICAN REP.
PUERTO RICO (US)

JAMAICA

HAITI

VIRGIN IS. (US)
ANTIGUA & BARBUDA

ST KITTS & NEVIS
ST VINCENT & GR.

GUADELOUPE (Fr)
DOMINICA
MARTINIQUE (Fr)

BELIZE

GUATEMALA
EL SALVADOR

HONDURAS

NICARAGUA

ARUBA(Neth.)

NETH. ANTILLES

ST LUCIA
BARBADOS

COSTA RICA

PANAMA

VENEZUELA

GRENADA

TRINIDAD & TOBAGO

GUYANA

SURINAME

FRENCH GUIANA

COLOMBIA

ECUADOR

PERU

BRAZIL

BOLIVIA

PARAGUAY

Rio de Janeiro
12.4m

CHI

URUGUAY

Buenos Aires
14.6m

ARGENTINA

MAURITAN

CAPE VERDE

SENE
GAMBIA

GUINEA-BISS

GU
SIERRA

Around 40 percent of the world's population lives less than 60 miles from the coast – within reach of severe coastal storms. About 100 million people live less than one meter above mean sea level. More people are gravitating to these areas of fast-growing economic development, but coastal erosion, rising sea levels, saltwater contamination, and potentially more powerful storms, are expected to put these already threatened environments under increasing stress.

Some of these consequences of climate change, such as the inundation of large delta areas, are potentially catastrophic. Others, such as the movement of saltwater upstream into freshwater rivers, will take their toll more slowly, as drinking and irrigation water becomes saline, river water becomes too corrosive to use for cooling in industrial processes and power plants, and changing coastal habitats affect wildlife.

While all coastal cities face such threats, the impact on those with over 10 million inhabitants will be most substantial. Water and sanitation systems may be placed under unbearable strain, and millions of poor people in shanty towns on the fringes of the cities may be at even greater risk from disease. Port facilities may no longer be viable, and government and financial services may be severely damaged, affecting the administration and economy of the entire country.

Cities at Risk

Coastal erosion, saltwater intrusion into freshwater supplies, and coastal storms all combine to threaten coastal areas – often regions of high population growth and intensive economic development.

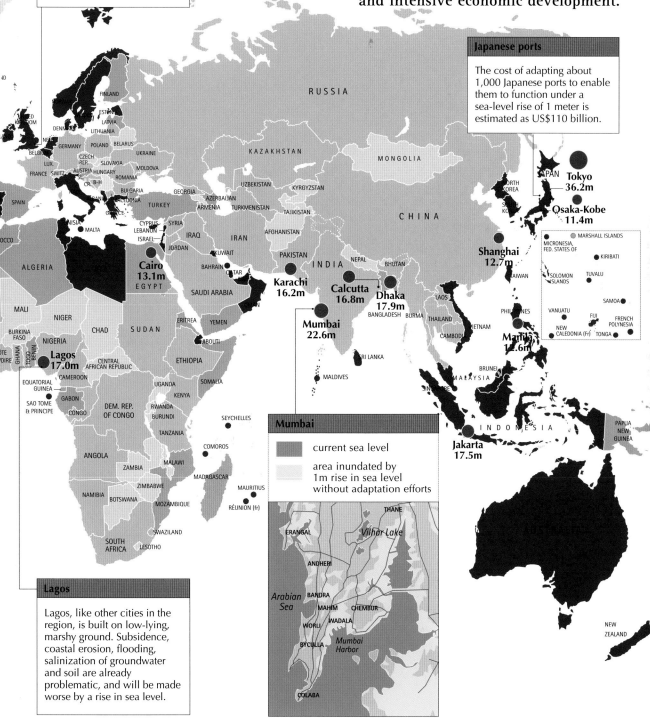

London

London is protected from particularly high tides and storm surges by the Thames Barrier, but more frequent storms and the pressure of rising sea-level increases the risk to the city.

Japanese ports

The cost of adapting about 1,000 Japanese ports to enable them to function under a sea-level rise of 1 meter is estimated as US$110 billion.

Lagos

Lagos, like other cities in the region, is built on low-lying, marshy ground. Subsidence, coastal erosion, flooding, salinization of groundwater and soil are already problematic, and will be made worse by a rise in sea level.

Mumbai

- current sea level
- area inundated by 1m rise in sea level without adaptation efforts

Tokyo 36.2m
Osaka-Kobe 11.4m
Shanghai 12.7m
Cairo 13.1m
Karachi 16.2m
Calcutta 16.8m
Dhaka 17.9m
Mumbai 22.6m
Manila 12.6m
Lagos 17.0m
Jakarta 17.5m

THREATENED HERITAGE

2005
selected sites

Climate change is threatening parts of the world's cultural and historical heritage, but the possible loss or damage to these irreplaceable sites is rarely represented in economic estimates of the cost of climate change.

There is no comprehensive inventory of important cultural and historical sites under threat from climate change. Many are located in coastal areas, where the threat of sea-level rise and retreating coastlines suggests, at the very least, a new set of conservation challenges. The potential for more frequent floods and more intense storms poses greater threats to cherished buildings, monuments, archaeological sites, and other material traces of history and heritage.

In other areas, changes in local climate may force adjustments in long-held cultural traditions and relegate to history some climate-related features that once reinforced shared values and memories.

Arctic

Indigenous people will find it increasingly difficult to maintain traditional hunting and fishing skills if the sea-ice melts, affecting seal and polar bear populations. A northward shift in vegetation zones will also take with it their other traditional food source: the tundra-grazing caribou and reindeer.

Boston, USA

The combination of sea-level rise and coastal storms could increase the height of flooding on the Charles River and inundate famous historical sites of Boston, such as Boston Garden.

Belize Barrier Reef

In 1842, Charles Darwin described it as "the most remarkable reef in the West Indies". Now a World Heritage Site, the reef has suffered "bleaching" due to higher water temperatures, and will suffer further if ocean temperatures rise as predicted.

Cultural Losses

Damage to indigenous cultures, historical monuments and archeological sites adds to the incalculable economic losses of climate change.

Scotland, UK

A survey found 12,000 sites vulnerable to coastal erosion, which is accelerated by sea-level rise. Archaeologists expect that 600 sites will be of exceptional importance.

Czech Republic

In 2002, flooding across Europe damaged concert halls, theatres, museums, and libraries. An estimated half a million books plus, in addition to archival documents, were damaged. Climate change may bring more flooding and further losses to some areas.

Mt. Everest

When Edmund Hillary returned to Everest in 2003 he remarked that, whereas in 1953 snow and ice had reached all the way to base camp, it now ended five miles away. Local people are worried that if Everest loses its natural beauty, tourists will stay away, destroying local livelihoods.

Thailand

Severe flooding currently threatens historical sites. In northeastern Thailand, floods have damaged the 600-year-old ruins of Sukothai, the country's first capital, and the ruins of Ayutthaya, which served as the capital from the 14th to 18th centuries.

Alexandria, Egypt

The monuments of Alexandria, including the 15th-century Qait Bey Citadel, are threatened by coastal erosion and the inundation of the Nile delta region, caused by sea-level rise.

Venice, Italy

The structural integrity of many buildings is being damaged by frequent flooding. St. Mark's Square floods about 50 more times each year than it did in the early 1900s. Current flooding is caused mostly by land subsidence, but rising sea levels exacerbate the problem. Engineering efforts are underway to protect the city.

Tuvalu

Leaders in Tuvalu are making plans for the evacuation of this island nation. It is unclear how communities, culture, and traditions will be maintained within another country.

West Coast National Park, South Africa

The oldest human footprints, estimated to have been made 117,000 years ago, were found near Langebaan Lagoon in the national park. To avoid vandalism, these fossils have been moved, but other pieces of history remain undiscovered and at risk of inundation because of sea-level rise.

Part 5   Responding to Change

The challenges of mitigating and adapting to climate changes are unprecedented, but not insurmountable. There are key elements and institutions to build upon; equally there is much to be done and delay will result in higher future costs. The scale of reduction in greenhouse gas emissions to safer levels requires truly international cooperation. A twin goal must be to facilitate adaptation among those least able to protect themselves from climate impacts.

Negotiations over climate change responses recognized a complex patterns of causes, relative contributions to the problem, ability to contribute to solutions, potential for benefits and losses, and irreversible changes. Laid over existing political differences, international trade negotiations, and other factors of international relations, the dialogue to develop and ratify the Kyoto Protocol and the first round of commitments did not move quickly.

While most nations have signed international agreements, they are also faced with forging agreements at home. Meeting reduction targets requires the involvement of local government, small businesses, corporations, and civil organizations; even religious organizations are taking a strong stand to reduce emissions. In nations that have not committed to the Kyoto Protocol, such as the USA and Australia, municipalities and companies have organized to reduce emissions and are challenging their national leaders.

The relationship between emissions and economic growth is not inevitable. The greater diversity of renewable energy technologies in more recent years opens more opportunities for energy consumption with low emissions. For developing countries, investing early in a less carbon-dependent infrastructure offers the potential for long-term savings.

Many companies have demonstrated the cost effectiveness of reducing energy use, switching fuels and controlling greenhouse gas emissions. Yet, not all technological adjustments will be easy. Decades of investment in road transportation and private vehicles has created a path dependence that will be expensive to alter. Carbon trading is growing but does not reduce emissions on its own.

International aid provides funding to support some of these efforts, but it is a very small fraction of development assistance. Supporting the capacity of local populations to cope with the challenges of current climatic hazards and longer-term climate change remains a priority. Linking energy efficiency and management of climatic risks with sustainable development, through education, health care, employment and information, is essential.goals.

"To the reticent nations, including the United States, I'd say this: there is such a thing as a global conscience and now is the time to listen to it."

Paul Martin, former Canadian Prime Minister, 2005

SIGNATORIES TO UNFCCC AND KYOTO

February 2006

███ signed the UNFCCC, but not adopted the Kyoto Protocol

███ adopted the Kyoto Protocol   ███ non-signatories

Selected formal or informal negotiating blocs:
—— Association of Small Island States (AOSIS)

⌐ Umbrella Group

Environmental Integrity Group (EIG)

● Organization of Petroleum Exporting Countries (OPEC)

The UN Framework Convention on Climate Change (UNFCCC) aims to stabilize greenhouse gas emissions "at a level that would prevent dangerous anthropogenic [human-induced] interference with the climate system".

This level should be achieved within a time-frame that allows ecosystems to adapt to climate change, ensures that food production is not threatened, and enables economic development to proceed in a "sustainable manner". Agreed in Rio in June 1992, the Convention came into force in March 1994.

The Convention places the initial onus on the industrialized nations and 12 economies in transition to reduce their emissions, and finance developing countries' search for strategies to limit their own emissions in ways that will not hinder their economic progress.

The Convention is a flexible framework, clearly recognizing that there is a problem. The first addition to the treaty, the Kyoto Protocol, set targets for reductions in emissions. Adopted in 1997, it came into force in February 2005. The USA and Australia have signed the Convention but not the Protocol, creating uncertainty around the next steps.

Climate change continues to be high on the international agenda, but there is still much disagreement as to what to do, when, by whom. Conflicts over relative responsibilities for reducing emissions and funding adaptation continue to slow down negotiations.

Observers

Examples of business, research and environmental groups participating as observers in negotiations:

Competitive Enterprise Institute: Funded by Exxon Mobil, it issues press releases exposing supposed flaws in climate science.

American Petroleum Institute: A trade association for the US oil industry, it represents the interests of the industry in the climate change debate.

International Institute for Sustainable Development: Publishes the Earth Negotiation Bulletin and provides independent reviews.

New Economics Foundation: An independent "think and do" tank that promotes innovative solutions that challenge mainstream thinking on economic, environment and social issues.

Climate Action Network: A worldwide network of over 340 NGOs with an active role in the process.

Generation Foundation: Co-founded by Al Gore, former US vice president, it supports sustainable development and sustainability research.

Greenpeace: Complements its highly visible public protests with science and lobbying.

International Action

Most countries have acknowledged the problem of climate change by signing the Convention on Climate Change.

Negotiating groups

Negotiation takes place within the Convention framework to develop policies and operational procedures. Countries tend to affiliate themselves to others with similar agendas. The main groupings are:

Alliance of Small Island States (AOSIS): a coalition of some 43 low-lying and small island countries that are particularly vulnerable to sea-level rise. See underlined names on map.

Developing countries: over 130 members (including China) of the Group of 77, forming a diverse group with differing interests on climate change issues. See list on http://www.g77.org/main/main.htm

European Union (EU): 25 members are parties to the convention, but EU presidency often voices their collective view. Members listed on: www.eurunion.org/states/home.htm

Environmental Integrity Group (EIG): a coalition that highlights environmental issues during negotiations and attempts to minimize the trade in carbon sinks. See map.

Least Developed Countries (LDCs): Countries defined as such by the UN. They work together, particularly on adaptation and capacity building. www.un.org/special-rep/ohrlls/ldc/list.htm

Organization of Petroleum Exporting Countries (OPEC): works on issues related to mitigation policy, including compensation for the adverse effects of reduced fossil fuel consumption. See map.

Umbrella Group: A loose coalition of non-EU developed countries that occasionally act together. See map.

In February 2005, seven years after the Kyoto Protocol was negotiated, the commitments agreed to became legally binding after countries representing at least 55 percent of greenhouse gas emissions by the industrialized world had ratified the protocol.

Although the Protocol had been ratified by over 160 countries by April 2006, only the Annex I countries – those already industrialized or whose economies were considered to be in transition in 1997 – had been set targets under the Protocol. This is because they have been emitting greenhouse gases into the atmosphere for a long period of time, and are considered able to afford reductions. These countries are committed to reducing their greenhouse gas emissions to a combined average of 5.2 percent below their 1990 levels by a target date between 2008 and 2012.

Under the Protocol, the industrialized countries were to have made "demonstrable progress" by 2005. Some are making progress towards their goals, while others have actually increased their emissions. The refusal of the USA and Australia to ratify the Protocol means that more than 40 percent of current emissions by Annex I countries are not covered under this agreement.

The reductions under the Kyoto Protocol are considered a first step, as EU Environment Ministers recommend that reductions of between 60 percent and 80 percent by 2050 will be needed to avert more serious climate change impacts. As a result of agreements reached in Montreal at the end of 2005, a working group was established to start discussing commitments to reduction after 2012.

Long-term effect of CO₂ emissions

Assuming a rapid decline
after 2050

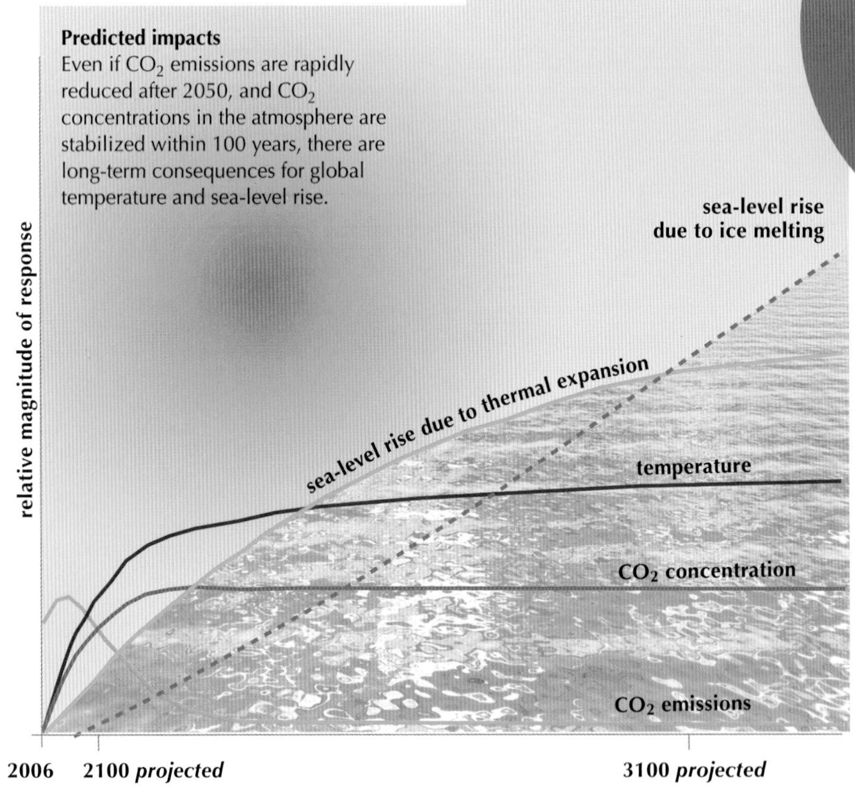

Predicted impacts
Even if CO_2 emissions are rapidly reduced after 2050, and CO_2 concentrations in the atmosphere are stabilized within 100 years, there are long-term consequences for global temperature and sea-level rise.

sea-level rise
due to ice melting

sea-level rise due to thermal expansion

temperature

CO_2 concentration

CO_2 emissions

relative magnitude of response

2006   2100 *projected*   3100 *projected*

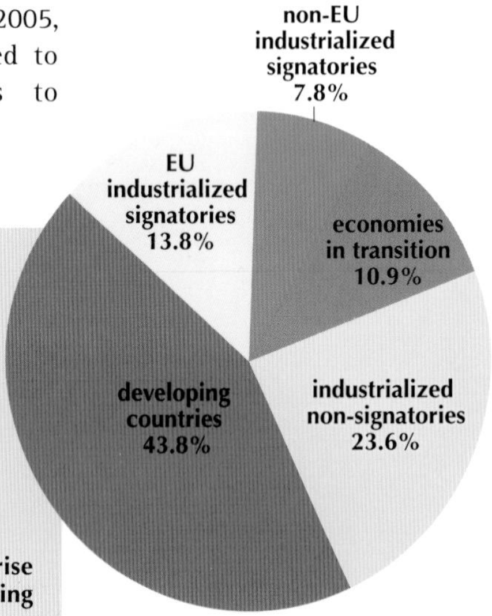

Greenhouse gas emissions with land-use change

Contribution to
annual global emission
by economic status
2003 or latest available data

non-EU industrialized signatories
7.8%

EU industrialized signatories
13.8%

economies in transition
10.9%

developing countries
43.8%

industrialized non-signatories
23.6%

Meeting Kyoto Targets

Many countries are making progress towards their Kyoto commitments, but even the agreed targets fall far short of stabilizing greenhouse gas emissions at levels considered to be safe.

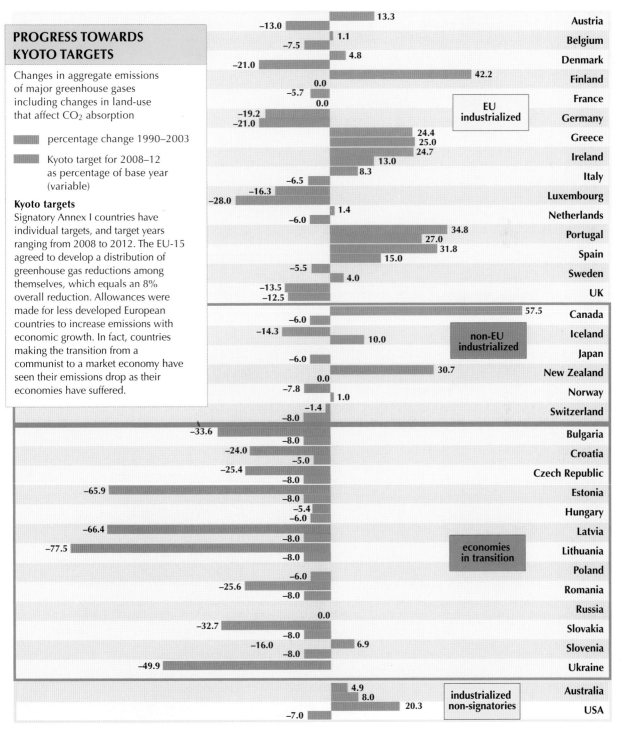

PROGRESS TOWARDS KYOTO TARGETS

Changes in aggregate emissions of major greenhouse gases including changes in land-use that affect CO_2 absorption

- percentage change 1990–2003
- Kyoto target for 2008–12 as percentage of base year (variable)

Kyoto targets
Signatory Annex I countries have individual targets, and target years ranging from 2008 to 2012. The EU-15 agreed to develop a distribution of greenhouse gas reductions among themselves, which equals an 8% overall reduction. Allowances were made for less developed European countries to increase emissions with economic growth. In fact, countries making the transition from a communist to a market economy have seen their emissions drop as their economies have suffered.

EU industrialized

Country	percentage change 1990–2003	Kyoto target
Austria	13.3	−13.0
Belgium	1.1	−7.5
Denmark	4.8	−21.0
Finland	42.2	0.0
France	−5.7	0.0
Germany	−19.2	−21.0
Greece	24.4	25.0
Ireland	24.7	13.0
Italy	8.3	−6.5
Luxembourg	−16.3	−28.0
Netherlands	1.4	−6.0
Portugal	34.8	27.0
Spain	31.8	15.0
Sweden	−5.5	4.0
UK	−13.5	−12.5

non-EU industrialized

Country	percentage change 1990–2003	Kyoto target
Canada	57.5	−6.0
Iceland	−14.3	10.0
Japan	−6.0	
New Zealand	30.7	0.0
Norway	−7.8	1.0
Switzerland	−1.4	−8.0

economies in transition

Country	percentage change 1990–2003	Kyoto target
Bulgaria	−33.6	−8.0
Croatia	−24.0	−5.0
Czech Republic	−25.4	−8.0
Estonia	−65.9	−8.0
Hungary	−5.4	−6.0
Latvia	−66.4	−8.0
Lithuania	−77.5	−8.0
Poland		−6.0
Romania	−25.6	−8.0
Russia	0.0	
Slovakia	−32.7	−8.0
Slovenia	−16.0	6.9
Ukraine	−49.9	−8.0

industrialized non-signatories

Country	percentage change 1990–2003	Kyoto target
Australia	4.9	8.0
USA	20.3	−7.0

Under the Kyoto Protocol, countries required to reduce their emissions are entitled to purchase "carbon credits" from developing countries, or from industrialized countries whose emissions are below the level required. The credits cover emissions of all greenhouse gases, expressed as carbon dioxide equivalents (CO_2e). The trade in carbon credits is intended to encourage investment in energy efficiency, renewable energy and other ways of reducing emissions.

Many industrialized countries have devolved this responsibility on to the commercial organizations that are creating the greenhouse gases, and much of the trade therefore takes place at a company level. Carbon markets facilitate the trade and there are two main types: project-based markets and allowance-based markets.

Project-based markets encourage investment in companies or schemes committed to reducing emissions. These are dominated by projects implemented as part of the Clean Development Mechanism (CDM) or Joint Implementation (JI). The main buyers are the industrialized and transition economies. The main sellers are in Asia and South America, with India and Brazil taking the lead. Africa seems to be missing out on the benefits of investment, however, with only one major project reported for 2004.

Allowance-based markets enable large companies, such as energy producers, to purchase emissions allowances under schemes administered by international bodies such as the EU Emissions Trading Scheme (ETS). These so-called "cap and trade" programmes cover a small proportion of trade, but the market is growing.

Carbon markets still account for less than 0.5 percent of annual global greenhouse gas emissions. They have also been criticized by environmentalists for including unsustainable projects, such as single-species plantations, large hydroelectric dams, and oil and coal production.

Nevertheless, carbon trading is an increasingly important element in international efforts to slow climate change. An encouraging sign is that markets have been set up in Australia and the USA, which have not adopted the Protocol, to facilitate emissions reduction. Placing a price on greenhouse gas emissions encourages innovation to reduce them.

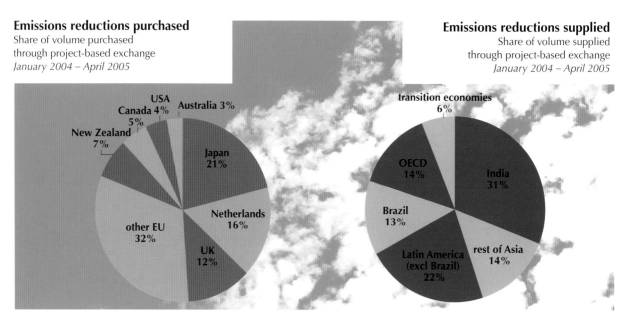

Emissions reductions purchased
Share of volume purchased
through project-based exchange
January 2004 – April 2005

USA Canada 4% Australia 3%
5%
New Zealand 7%
Japan 21%
Netherlands 16%
other EU 32%
UK 12%

Emissions reductions supplied
Share of volume supplied
through project-based exchange
January 2004 – April 2005

transition economies 6%
OECD 14%
India 31%
Brazil 13%
Latin America (excl Brazil) 22%
rest of Asia 14%

Carbon Trading

Trading in carbon credits is one way to share the burden of reducing emissions globally.

Project-based exchange

Volume exchanged
1998–2004
million tonnes CO_2e

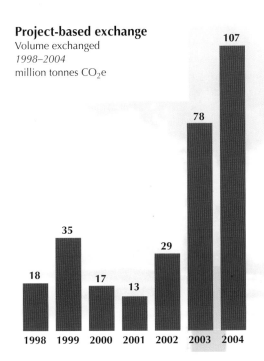

1998	1999	2000	2001	2002	2003	2004
18	35	17	13	29	78	107

Types of projects

Technology share of
emission-reduction projects
2004–05
percentage of total CO_2e contracted

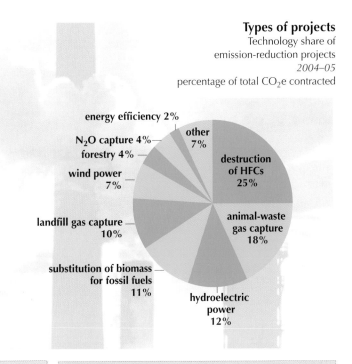

- energy efficiency 2%
- N₂O capture 4%
- forestry 4%
- wind power 7%
- landfill gas capture 10%
- substitution of biomass for fossil fuels 11%
- other 7%
- destruction of HFCs 25%
- animal-waste gas capture 18%
- hydroelectric power 12%

Project-based market

Buyers invest in projects or companies where there is a commitment to use the money to reduce greenhouse gas emissions, compared with what would have happened without the investment. Buyers (countries, companies or individuals) are awarded "carbon credits" in exchange. These can be used to meet regulatory targets for greenhouse gas emissions, such as those set in the Kyoto Protocol, or voluntary targets that companies are increasingly setting for themselves.

Examples of projects financed under such schemes include a sugar factory in India powered by renewable resources, reforestation in Costa Rica, and a scheme to provide photovoltaic-powered ICT systems in developing countries.

In Honduras, a 12.7 megawatt hydroelectric plant, powered by river run-off, has been supported by the sale of 310,000 tonnes of CO_2 emissions reductions through the World Bank's Community Development Carbon Fund.

In Colombia, 15 wind turbines have been funded to provide power for the indigenous Wayuu people. The wind turbines will save more than 1.1 million tonnes of CO_2 over 21 years, compared with emissions from fossil-fueled power.

Allowance-based exchange

Volume exchanged
2002–05
million tonnes of CO_2 equivalent
estimates

Allowance-based markets enable companies to offset their emissions by purchasing credits from countries that either have no limit placed on their emissions, or have kept emissions below the level required. Trade has grown rapidly since 2003, boosted by the opening of the EU Emissions Trading Scheme in January 2005. Schemes are also operating in Australia, Canada and the USA, including the Chicago Climate Exchange (CCX), established by a number of large corporations and the World Resources Institute.

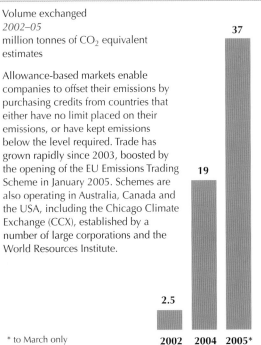

2002	2004	2005*
2.5	19	37

* to March only

Major adaptation funds

The following funds have been established by the Global Environment Facility (GEF), as agreed under the Convention on Climate Change:
- The Least Developed Country Fund (LDCF) supports the development and implementation of National Adaptation Plans of Action (NAPA) for the least developed countries.
- The Special Climate Change Fund (SCCF) addresses the special needs of developing countries.
- The Adaptation Fund (AF) will use the 2% surcharge on Clean Development Mechanism (CDM) projects for adaptation.
- Strategic Priority: Piloting an Operational Approach to Adaptation (SPA) supports projects combining mitigation and adaptation.

Developing countries need money to help them meet the challenges of climate change, but the amount available is a minuscule proportion of overseas development assistance.

Funding agreed under the Convention on Climate Change is distributed mainly through the Global Environment Facility (GEF), which supports action in developing countries on a range of environmental issues. Countries were initially helped to assess their emissions, but the focus is now on reducing emissions, and on adapting to change.

The regional development banks and other donors are increasingly donating money to climate-response projects, and the Asia–Pacific Partnership of USA, Australia, China, India, Japan and South Korea is seeking, and investing in, technological solutions in developing countries.

The private sector is also getting involved. Companies of all sizes are taking stock of their risks and working out how to protect the value of their investments by minimizing the impact of climate change.

There might be more funding for both mitigation and adaptation if developing countries, which have contributed little to climate change but are likely to suffer increasingly severe impacts, were to make a successful case for compensation from the major greenhouse gas emitters.

A drop in the ocean
Total GEF spending on climate change projects as percentage of total official and private overseas development aid *1991–2004*

GEF spending on climate change: $1,849 million

development aid: $2,697,820 million

Argentina: Renewable energy

$13.6m

As part of a rural electrification program, photovoltaics, wind and mini-hydro schemes are being substituted for diesel generators, supplying power to over 130,000 rural households and about 4,400 public services.

Financing the Response

Current funding is inadequate to help countries respond to climate change.

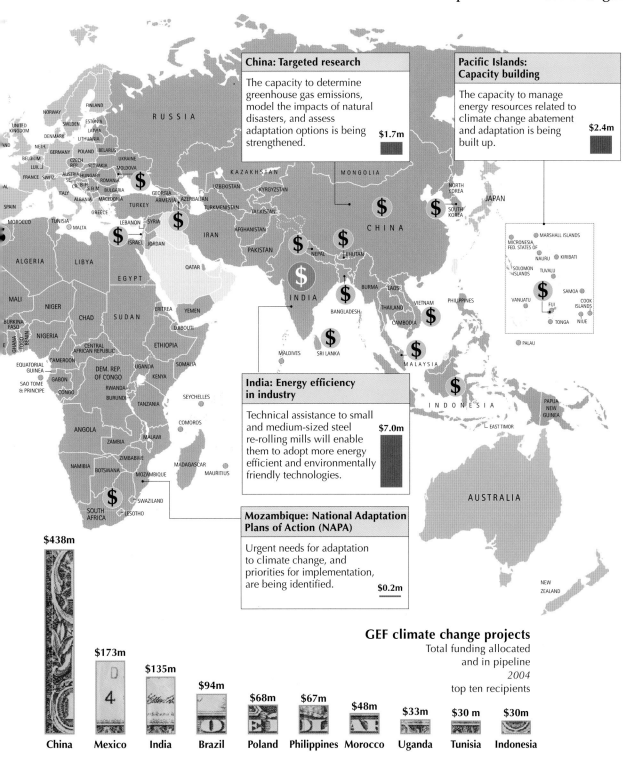

China: Targeted research

The capacity to determine greenhouse gas emissions, model the impacts of natural disasters, and assess adaptation options is being strengthened.

$1.7m

Pacific Islands: Capacity building

The capacity to manage energy resources related to climate change abatement and adaptation is being built up.

$2.4m

India: Energy efficiency in industry

Technical assistance to small and medium-sized steel re-rolling mills will enable them to adopt more energy efficient and environmentally friendly technologies.

$7.0m

Mozambique: National Adaptation Plans of Action (NAPA)

Urgent needs for adaptation to climate change, and priorities for implementation, are being identified.

$0.2m

GEF climate change projects
Total funding allocated and in pipeline
2004
top ten recipients

$438m	$173m	$135m	$94m	$68m	$67m	$48m	$33m	$30 m	$30m
China	Mexico	India	Brazil	Poland	Philippines	Morocco	Uganda	Tunisia	Indonesia

CITIES COMMITTED TO CHANGE

Number of cities or local governments participating in Cities for Climate Protection *2005*

- ■ 100 or more
- ■ 21 – 50
- ■ 11 – 20
- ▨ 6 – 10
- ▨ 1 – 5
- ▨ not a participant

Cities around the world are not waiting for national governments to debate the implementation of the Climate Change Convention. They have signed their own commitment to reducing greenhouse-gas emissions, as part of the campaign Cities for Climate Protection (CCP).

The campaign, established in 1993 by the International Council for Local Environmental Initiatives (ICLEI), has engaged over 670 local and city governments, which between them are responsible for an estimated 8 percent of the world's carbon dioxide (CO_2) emissions.

In the USA, which has not signed the Kyoto Protocol, concern has led to local and regional action. Mayors issued a statement in 2003, urging the national government to slow the rate of global warming. In February 2005, as the Kyoto Protocol came into effect, the mayor of Seattle issued a Climate Protection Agreement, pledging to curb greenhouse gas emissions at a local level. The agreement was endorsed by the US Congress of Mayors and by May 2006, 230 mayors had signed up.

> We know the science.
> We see the threat...
> The time for action
> is now.
>
> Arnold Schwarzenegger
> Governor of California
> 1 June 2005

Regional Greenhouse Gas Initiative (RGGI)

■ participating states

The Regional Greenhouse Gas Initiative (RGGI) aims for a 10% reduction in carbon dioxide emissions from power plants by 2018 within the areas under its jurisdiction. It plans to operate a "cap and trade" system that sets limits, but allows participating states to trade emissions.

USA

The Mayors' Climate Protection Agreement 2005 includes the following:
- We urge the US Congress to pass bipartisan greenhouse gas reduction legislation...
- We urge the federal government and state governments to enact policies and programs to meet or beat the target of reducing global warming pollution levels to 7% below 1990 levels by 2012...
- We will strive to meet or exceed Kyoto Protocol targets for reducing global warming pollution by taking actions in our own operations and communities...

CCP Five-Milestone Framework

A commitment to:
- conduct an energy and emissions inventory and forecast
- establish an emissions target
- develop a Local Action Plan
- implement policies and measures to reduce emissions
- monitor and verify the results.

Adopted by local authorities joining the CCP campaign.

As well as a reduction in greenhouse gas emissions, local residents may benefit from:
- financial savings through energy and fuel efficiency
- economic development and job creation
- preservation of green spaces
- reduction in air pollution
- reduction in traffic congestion.

Curitiba, Brazil

The city's urban planning is fostering more efficient public transportation.

Local Commitment

In many places, local and regional authorities are developing more aggressive emission reduction policies than federal governments.

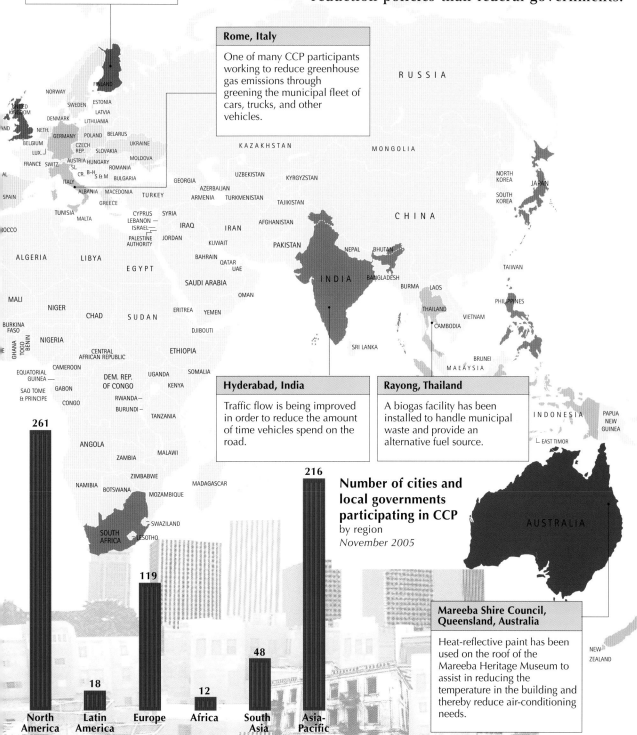

Finland

Cities participating in Finland represent almost 50% of the national population.

Rome, Italy

One of many CCP participants working to reduce greenhouse gas emissions through greening the municipal fleet of cars, trucks, and other vehicles.

Hyderabad, India

Traffic flow is being improved in order to reduce the amount of time vehicles spend on the road.

Rayong, Thailand

A biogas facility has been installed to handle municipal waste and provide an alternative fuel source.

Number of cities and local governments participating in CCP
by region
November 2005

Mareeba Shire Council, Queensland, Australia

Heat-reflective paint has been used on the roof of the Mareeba Heritage Museum to assist in reducing the temperature in the building and thereby reduce air-conditioning needs.

Region	Number
North America	261
Latin America	18
Europe	119
Africa	12
South Asia	48
Asia-Pacific	216

A country's "carbon intensity" is a measure of the efficiency of its economic output with respect to its carbon dioxide (CO_2) emissions. Countries with a low carbon intensity release relatively little carbon dioxide into the atmosphere when compared with their economic output. Their economies are considered to be comparatively "clean".

Industrialization has tended initially to develop through industries with high carbon dioxide emissions, such as shipping, steel and manufacturing. Only as an economy matured, with the growth of hi-tech industries, and the use of more efficient technology to process natural resources, has high economic output become associated with less pollution.

This historic pathway need not be taken by newly industrializing countries, however. Economic growth can be achieved with lower greenhouse gas emissions. A growing awareness about greenhouse gases, and the implementation of policies to force corporations to be environmentally responsive are essential. Emerging economies such as India and China, with carbon intensities five and seven times that of the UK, need to find ways of breaking the current link between high emissions and economic growth. China is currently being pressed to make energy and infrastructure investment, as its economy is set to quadruple in size by 2020.

If emissions are to be reduced even while economies grow, more efficient technology needs to be introduced. A key policy to achieve this is the Clean Development Mechanism, part of the Kyoto Protocol, which increases foreign investment in efficient technologies in emerging economies.

> There is an urgent need to consider ways to accelerate the decoupling of energy and CO_2 emissions from economic growth.
>
> Claude Mandil,
> International
> Energy Agency,
> *2004*

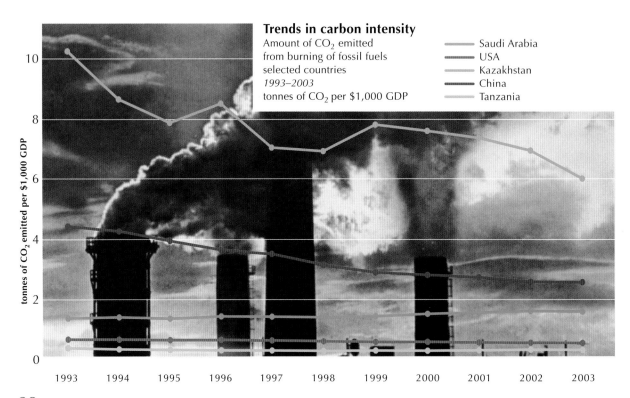

Trends in carbon intensity
Amount of CO_2 emitted from burning of fossil fuels selected countries
1993–2003
tonnes of CO_2 per \$1,000 GDP

- Saudi Arabia
- USA
- Kazakhstan
- China
- Tanzania

y-axis: tonnes of CO_2 emitted per \$1,000 GDP — 0, 2, 4, 6, 8, 10

x-axis: 1993, 1994, 1995, 1996, 1997, 1998, 1999, 2000, 2001, 2002, 2003

Carbon Dioxide and Economic Growth

Economic growth can be achieved with lower greenhouse gas emissions.

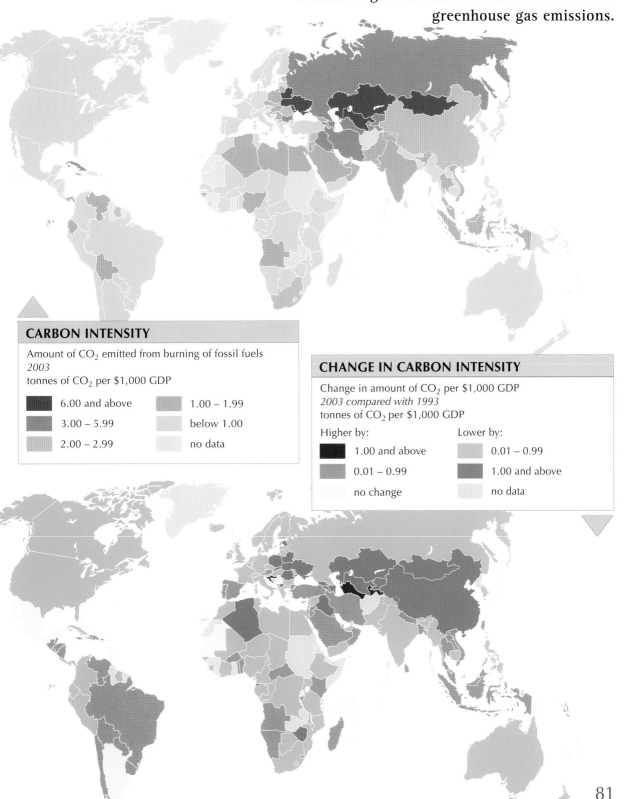

CARBON INTENSITY

Amount of CO_2 emitted from burning of fossil fuels
2003
tonnes of CO_2 per $1,000 GDP

- 6.00 and above
- 3.00 – 5.99
- 2.00 – 2.99
- 1.00 – 1.99
- below 1.00
- no data

CHANGE IN CARBON INTENSITY

Change in amount of CO_2 per $1,000 GDP
2003 compared with 1993
tonnes of CO_2 per $1,000 GDP

Higher by:
- 1.00 and above
- 0.01 – 0.99
- no change

Lower by:
- 0.01 – 0.99
- 1.00 and above
- no data

Increasing the use of renewable energy sources is an important way of reducing greenhouse gas emissions, while continuing to provide power. There is a tremendous diversity of renewable energy technologies, such as geothermal, solar thermal, solar photovoltaic, tide, and wind. However, even in countries with ample resources, renewable sources represent only a small percentage of total primary energy supply.

Worldwide, only about 4 percent of total primary energy comes from sources which do not produce greenhouse gases. Although this estimate under represents small-scale energy systems, it is clear that renewable sources could make a much larger contribution to the annual world power production of 116 million gigawatt hours.

Investment in renewable energy technologies is increasing rapidly, however. By 2035, the International Energy Agency expects investment in renewables by OECD countries to reach 30 percent of all investment in new power generation. An increasing number of countries, provinces and states are enacting legislation requiring utility companies to purchase a proportion of energy from renewable sources, or ensuring favorable prices to producers of renewable energy.

During 2004, the energy generated by wind increased by 20 percent worldwide. With over 50 countries participating in the Global Wind Energy Council, giant turbines are becoming part of the landscape.

Photovoltaics, a technology that was under developed until the 1990s, is now expanding rapidly. These solar cells are flexible enough to provide electricity for rural communities not linked to a large electrical grid, and, where possible, to contribute "green" energy to the broader network.

An estimated 40 million households, 60 percent of them in China, use solar thermal energy to heat water. Two million buildings in 30 countries use geothermal heat pump technology. The increasingly popular biodiesel fuels also offer significant reductions in emissions of greenhouse gas.

Solar energy
Production of photovoltaic cells
1996–2004
megawatts

- Japan
- Europe
- USA
- rest of world

	1996	2000	2004
Japan	21	129	602
Europe	19	61	314
USA	39	75	139
rest of world	10	23	140

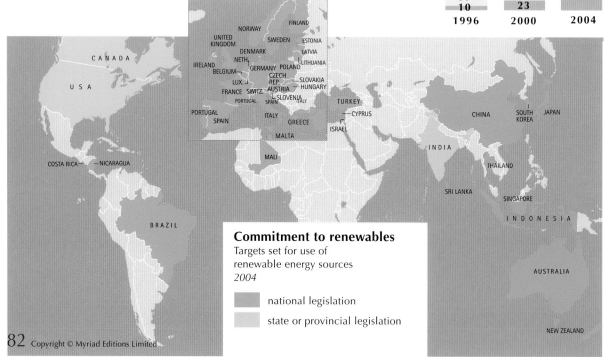

Commitment to renewables
Targets set for use of renewable energy sources
2004

- national legislation
- state or provincial legislation

Renewable Energy

Renewable energy could be the technological key to economically and socially sustainable societies.

Renewable energy sources
2004

■ electricity produced, GW installed capacity

▨ heat produced, GWth installed capacity

Small-scale hydro

Hydro systems generate electricity from running water. They can provide power for isolated villages, or feed power into the electricity grid. Small-scale hydro does not involve artificial reservoirs, and so avoids the formation and release of methane from decaying biomass.

61

Wind

Wind turbines of varying sizes are used to generate electricity, for the national grid or for isolated communities.

48

Biomass

Plant material – purpose grown or waste – can be burned or fermented, and used to generate electricity or heat. The CO_2 released is the same amount as was removed from the atmosphere during the plant's lifetime, so biomass is considered carbon neutral.

220

39

Geothermal

In geologically active areas, the Earth's intense heat can fuel power plants. Elsewhere, its temperature, which remains constant 1.5 meters below the surface, can be used to heat and cool buildings.

28

8.9

Solar

Photovoltaic panels convert the sun's radiation into electricity.

Thermal panels convert the sun's radiation into heat.

77

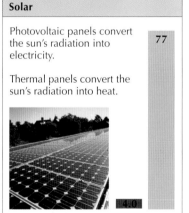

4.0

Tide, wave, ocean

The movement of the sea can be used to generate electricity.

0.3

Renewable energy production

Share of production by greenhouse gas neutral technologies for selected countries
2003
gigawatt hours

▨ geothermal
▨ solar
▨ wind
▨ biomass

Australia
1,367
0.4%
48.5% 51.1%

UK
4,907
0.1%
26.2%
73.8%

Japan
5,259
0.04%
17.9%
15.8%
66.3%

Spain
12,874
0.3%
5.9%
93.8%

Germany
22,274
0.5%
1.5%
13.3%
84.7%

USA
33,107
19.3%
44.9%
34.1%
1.7%

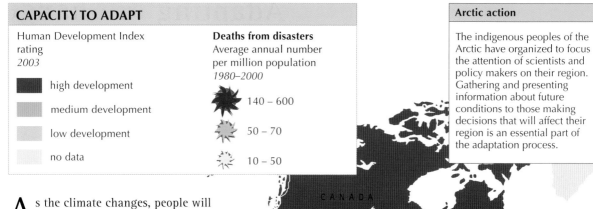

CAPACITY TO ADAPT

Human Development Index
rating
2003

▇ high development

▇ medium development

▇ low development

▇ no data

Deaths from disasters
Average annual number
per million population
1980–2000

✦ 140 – 600

✦ 50 – 70

✦ 10 – 50

Arctic action

The indigenous peoples of the
Arctic have organized to focus
the attention of scientists and
policy makers on their region.
Gathering and presenting
information about future
conditions to those making
decisions that will affect their
region is an essential part of
the adaptation process.

As the climate changes, people will be forced to adapt. A broad measure of their capacity to do this is the Human Development Index (HDI). It scores countries according to their income, social equality, and provision made for education and health. Countries with higher scores are more likely to have the resources to invest in climate change adaptation.

The most urgent need is to cope with extreme weather events, which have affected countries of every level of development in recent years. The number of people affected continues to rise, although relatively fewer people are dying as countries adopt early warning systems, evacuation plans and shelters. However, disasters can still exceed the ability of poorer countries and people to cope.

Developing countries are not alone in struggling to cope with weather disasters. Some protections are imperfect and events may surprise the unprepared. While the risks to New Orleans were well known, Hurricane Katrina in 2005 took thousands of lives and caused damage of at least $50 billion. The heatwave that struck Europe in 2003 took around 35,000 lives, mostly among the elderly.

The building and maintenance of a sound public health infrastructure, necessary to face climate change, has social and economic benefits in years without disasters. Earlier climate change interventions are likely to result in lower long-term costs.

Adapting to Change

The capacity to adapt to climatic hazards and
stresses depends on a country's wealth,
resources and governance.

Bangladesh

The Bhola cyclone in 1970
killed over 300,000 people,
but by the time a similar
cyclone struck in 1991, the
Bangladesh government,
helped by NGOs, had already
embarked on the building of
communal cyclone shelters.
Even so 139,000 people died.
Since then, although cyclones
hit the coast annually, because
of better warning systems,
cyclone shelters and other
measures, no event has caused
more than 600 deaths.

Africa

The Famine Early Warning
System, which represents
decades of institutional and
scientific development, has
helped anticipate, and thereby
reduce harm from, droughts
across Sub-Saharan Africa. In
2006, however, millions of
people in the Horn of Africa
were still facing starvation.

STOP CO2
AMICI DELLA TERRA
STOP AI GAS SERRA
AMICI DELLA TERRA
AMICI DELLA TERRA
2050: TOURIST IN VENICE
2050: VENICE UNDER WATER
SAVE THE CLIMATE
AMICI DELLA TERRA
AMICI DELLA TERRA
FRIENDS OF THE EARTH
AMICOS DE LA TIERRA
PIAGET

PART 6 COMMITTING TO SOLUTIONS

Almost every day brings new evidence of the greenhouse effect: rises in average temperatures and sea-level, melting glaciers and ice shelves, consequences for plant and animal species, changes in people's lives.

The remaining uncertainties no longer need to delay concerted actions. While we cannot forecast with precision the timing, location, and intensity of the consequences of climate change, it is clear that many impacts will be severe and some will result in disasters with tremendous economic losses and incalculable human costs. Already, a Pacific island has been permanently evacuated, the infrastructure of polar regions is falling apart as the ground beneath it melts and softens, and the 2003 European heatwave has shown that no one is immune.

Delay comes with substantial risks. The climate system responds slowly, so the positive benefits of actions taken today will not be realized for decades. The longer we delay, the greater the risks and the more difficult it becomes to stabilize climate change at safe levels. Tipping points become more likely: that is, when the system shifts into entirely new states – like the thawing of permafrost and release of methane or the collapse of ice sheets resulting in tens of meters of sea-level rise.

Reducing greenhouse gas emissions can never make the climate and ecosystems return to the way they were. We are committed to some changes. Some of the worst consequences can be avoided, but only if we achieve much deeper reductions in emissions than have been internationally agreed so far. The first round of the Kyoto Protocol committed the industrialized countries among the signatories to an average 5 percent reduction. However, the European Union environment ministers estimate that a reduction of between 60 and 80 percent will be needed to prevent dangerous climate change.

Reducing emissions to this extent will require massive changes to the world's carbon-based economy and our current inefficient use of energy. The good news is that many of the required technologies, such as geothermal, solar and wind power, already exist, and there are many opportunities to improve and expand on their use.

Individual efforts are an essential starting point, and knowledge and technologies exist to support substantial improvements. But, meeting the scale of this challenge will require more than our personal commitment. The major changes in the design of efficient buildings, transportation, energy and other systems will require long-term vision, leadership, cooperation, innovation, and investment from business and governments at all levels. Assuring a public commitment to action is as important as our individual efforts.

"The way we use fuel and energy is causing global warming, but we can slow it down if we take action now. Global warming isn't cool. Stopping it is."

Kevin Bacon, actor, director, musician, 2005

Everybody can take action to combat climate change, and there is plenty of advice on how to do it. From carbon footprint calculators that allow us to evaluate our total carbon dioxide (CO_2) emissions, to specific actions we can take to reduce emissions, we can all become more conscientious about our use of resources. In addition to the climate benefits, most of the recommended actions result in long-term household financial savings, and many will lead to improved personal health and quality of life

Together, the actions of millions of people could add up to considerable savings in greenhouse gases, but they will not, on their own, be sufficient to halt climate change. Individuals also need to put pressure on government representatives and companies to take the larger-scale collective action necessary to achieve a reduction in emissions of 60 to 80 percent.

Take the One-Tonne Challenge
Reduce your annual carbon emissions by one tonne.
Go to: **www.climatechange.gc.ca** and follow the links.

Energy savings at home...

Using less energy not only helps the planet, but also saves money on household bills.

Turning the heating thermostat down, and the air conditioning up, by 1.5°C (3°F) saves around 1 tonne of CO_2 a year.

An energy-efficient refrigerator could save nearly half a tonne of CO_2 a year, compared with an older model.

3.0

Insulating windows, doors, and electrical outlets, and adding more insulation to the attic and basement reduces energy consumption.

Compact fluorescent, spiral light bulbs are 75% more efficient than standard light bulbs.

Locally produced, seasonal foods save the emissions resulting from transporting food long distances, or from heating greenhouses to grow out-of-season produce.

Energy savings on the road...

Walking, cycling, using a car pool or taking public transport, all produce fewer emissions than those emitted by a single person in a car.

Choosing the most efficient car available, such as a hybrid gasoline–electric model, and keeping any car well maintained, will reduce emissions.

Keeping tires optimally inflated uses less fuel and cuts down emissions.

Driving at 5 mph below the speed limit over an 8-mile commute to work saves 250 kg of CO_2 per year.

Sharing a car, and avoiding short journeys by car, saves energy.

Personal Action

People all over the world are taking measures to reduce the greenhouse gases emitted as a result of the way they live.

Comparative emissions of sample journeys
Annual emissions of CO_2 tonnes

Monthly business trip of 400 miles:

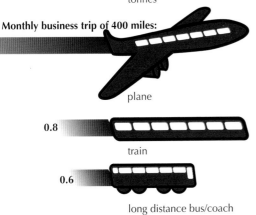

plane

0.8

train

0.6

long distance bus/coach

Daily commute of 10-mile round trip

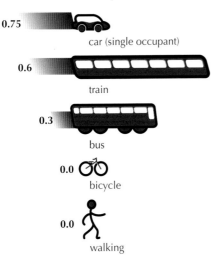

0.75

car (single occupant)

0.6

train

0.3

bus

0.0

bicycle

0.0

walking

Reducing garbage

On average a person throws away 10 times his or her bodyweight in rubbish per year. One kilogram sent to a landfill produces 2 kg of methane. The simplest way of reducing this burden is to buy and waste less unecessary packaging.

Recycling paper, glass, aluminum, steel and other materials to produce "new" materials, can make energy savings.

Using both sides of the paper and recycling it can save 2.5 kg of greenhouse gases for every kilogram of paper used.

Becoming carbon neutral

After reducing emissions as much as possible, people become carbon neutral by "offsetting" the rest.

They purchase "carbon credits" to channel their money into projects leading to a reduction in emissions. With details of the activity or fuel use to be offset, the organization calculates how many carbon credits need to be bought. Cost of credits varies, but is around $10/£7.50 per tonne of CO_2.

Buying green electricity can dramatically lower domestic emissions. Even where this is not yet an option, consumers in Europe, Japan, the USA, and other countries can still offset their electricity use by purchasing renewable energy certificates. The money generated goes to develop new and existing renewable energy facilities

Emissions calculators

Calculate your emissions at:

www.climatecare.org
www.earthfuture.com/climate/calculators/
www.carbonfund.org/
www.chooseclimate.org/flying/mapcalc.html
www.climatesolutions.org/

Climate change presents a central challenge: how to shift from the current path of social, economic and technological development to one that reduces greenhouse gas emissions and prepares us for the climates of the future. The necessary reduction in emissions of between 60 and 80 percent requires large-scale investment, policy development, and implementation. Although control of these areas rests largely with governments, businesses and civic organizations, it is the responsibility of individuals to leave the leaders of these institutions in no doubt of the need to take action.

Major corporations and government leaders at all levels have already brought about substantial reductions in greenhouse gas emissions, and made timely adaptations to climate change – in some cases in response to citizen and shareholder calls for action. Many of the adjustments made have provided substantial economic and other benefits to the companies and administrations involved. The growing number of companies and communities that have benefited from reducing greenhouse gas emissions demonstrates that there are opportunities to benefit while responding successfully to the challenges of reducing greenhouse gas emissions.

Advocating change

Individuals can encourage larger communities to act on climate change. For example:

- Workplaces and schools: encourage co-workers or fellow students to adopt strategies that reduce emissions.
- Companies and governing bodies: lobby management to invest in energy conservation measures, or renewable energy.
- Pressure groups and local government representatives: advocate local action. See Cities for Climate Protection program: www.iclei.org/co2/
- Corporations: encourage evaluation of their contributions to the greenhouse effect and point them to the many success stories and available toolkits.
- Government: lobby ministers to take actions to reduce emissions and plan adaptation options.

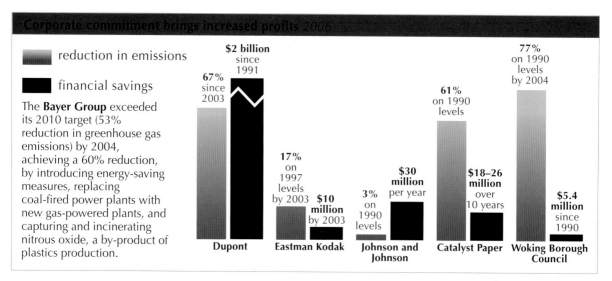

Corporate commitment brings increased profits 2006

- reduction in emissions
- financial savings

The **Bayer Group** exceeded its 2010 target (53% reduction in greenhouse gas emissions) by 2004, achieving a 60% reduction, by introducing energy-saving measures, replacing coal-fired power plants with new gas-powered plants, and capturing and incinerating nitrous oxide, a by-product of plastics production.

Dupont
67% since 2003
$2 billion since 1991

Eastman Kodak
17% on 1997 levels by 2003
$10 million by 2003

Johnson and Johnson
3% on 1990 levels
$30 million per year

Catalyst Paper
61% on 1990 levels
$18–26 million over 10 years

Woking Borough Council
77% on 1990 levels by 2004
$5.4 million since 1990

Public Action

The policies, practices, and investments of governments, businesses, and civic organizations will have the greatest impact on our future.

Online Tools ⟨ ⟩ ⊖ ⊕ ● ● ● ○

A growing number of websites are being set up to help organizations develop strategies to deal with climate change and reduce their emissions. These include:

- Climate Risk Toolkit for Corporate Leaders
  Addresses strategic and financial challenges of climate change:
  www.ceres.org
- UK Carbon Trust
  Support for businesses to save energy and develop carbon management plans: www.thecarbontrust.co.uk/carbontrust
- Clean Air Cool Planet's Campus Climate Action Toolkit for colleges and universities: www.cleanair-coolplanet.org/toolkit/
- USA's Environment Protection Agency's State Action Plans
  One set of models for larger areas:
  http://yosemite.epa.gov/oar/globalwarming.nsf/content/index.html
- Climate Action Registry Report On-line Tool (CARROT)
  Helping companies monitor emission reduction goals:
  https://www.climateregistry.org
- US Green Building Council's Leadership in Energy and Environmental Design (LEED) www.usgbc.org
  Criteria for efficient building construction and renovation, being adopted widely by the US military, and others.
- International Council on Local Environmental Initiatives (ICLEI)
  Technical support and toolkits for participants in the Cities for Climate Protection campaign: www.iclei.org
- Various business leadership groups offer strategies and case studies:
  www.theclimategroup.org; www.pewclimate.org
- Greenbiz offers links to toolkits for specific industries or goals:
  www.greenbiz.com/toolbox

Public debate on climate change – and what to do about it – is vigorously pursued through many different media.

The Adaptation Wizard

Even rapid reductions in greenhouse gas emissions will not prevent some climate change. Decision makers have begun to prepare for increased climate threats and changes in climatic resources, for instance by protecting flood hazard zones, designing buildings to reduce heat stress and conserving water resources. Everyone will need to adapt, from individuals to companies and governments.

Support for designing adaptation strategies:
www.ukcip.org.uk/resources/tools/adapt.asp

Home | Quantifying risks | Using this wizard | Decision-making and action plans | Principles of good climate adaptation | Adaptation strategy review | Resources

Principles of good climate adaptation

Here we describe some of the principles of good climate adaptation. You will apply these principles as you work your way through the wizard. Many of these are principles for good decision-making more generally.

- Work in partnership
- Keep a handle on uncertainty
- Frame your objectives carefully before you start
- Take a balanced approach to managing climate and non-climate risks
- Focus on actions to manage priority climate risks
- Use adaptive management to cope with uncertainty
- Try to find no-regret adaptation options
- Try to find win - win options
- Avoid actions that will make it more difficult to cope with climate risks
- Review your adaptation strategy regularly

PART 7 CLIMATE CHANGE DATA

U nderstanding patterns and trends, and anticipating exposure and impacts, both rely on global datasets developed over decades. Although this information is fundamental, it is surprisingly difficult to gather it, assure its quality and usefulness, and make it widely available and easily accessible.

The data presented in the maps and overleaf come from several different sources. Much of the data is self-reported by countries using agreed methods of accounting. But some countries do not have long monitoring records, so it is difficult to discern trends; others, for various reasons, do not report regularly. Since not all countries have equal resources for data collection and analysis, certain information is more reliable than others. For instance, it is easier and less expensive to identify a few major projects than many small household investments in renewable energy, so data are more likely to represent large-scale hydropower projects than to capture all small solar thermal water-heating systems or household improvements in energy conservation.

Another important challenge is that most of these data sets are reported at the national level. The causes of climate change, potential impacts, and the susceptibility of populations and environmental systems vary greatly within vast countries such as Russia, China, and the USA, and even within smaller countries like Ghana. For instance, population movements make it difficult to estimate how many people living in a coastal area might be affected by sea-level rise or severe storms; or how many living in water-stressed regions are likely to experience greater water stress.

Several efforts are underway to improve data. The UN Framework Convention on Climate Change (UNFCCC) and the Intergovernmental Panel on Climate Change (IPCC) reporting requirements are gathering more country data on key climate change related issues. There is a shared protocol for calculations to improve the consistency across national reports. The IPCC's *Assessment Report Four*, in three main volumes with associated summaries and syntheses, is due to be released in early 2007. The World Resources Institute has developed a useful web source for climate change related data – the Climate Analysis Indicators Tool (CAIT) – and other important information is distributed across a number of websites and printed reports.

Access to both web-based resources and printed materials remains a challenge for many scientists, decision-makers, and climate change negotiators in developing countries. Computer and web access is often limited and slow, but the data sets are easier to work with in digital form. Publications are expensive or difficult to obtain and their data more difficult to analyze. Making fundamental information more widely and easily available is essential to managing climate change risks.

"Cutting greenhouse gases is as optional as breathing."

Andrew Simms,
New Economics
Foundation, 2001

Countries	1 Population thousands 2004	2 GNI per capita US$ 2004	3 Human Development Index rating 2003	4 Water withdrawn as % of renewable water resources 2002 or latest available data	5 Coastal population as % of total population 1995	6 Weather-related disasters	
						number 2000–05	average annu deaths per million peop 1980–200(
Afghanistan	–	–	–	40%	0%	29	25
Albania	3,188	2,080	0.780	3%	97%	6	–
Algeria	32,373	2,280	0.722	35%	69%	20	1
Angola	13,963	1,030	0.445	0%	29%	15	0
Antigua and Barbuda	80	10,000	0.797	–	100%	–	–
Argentina	38,226	3,720	0.863	3%	45%	26	0
Armenia	3,050	1,120	0.759	27%	0%	2	–
Australia	20,120	26,900	0.955	3%	90%	39	1
Austria	8,115	32,300	0.936	3%	2%	8	–
Azerbaijan	8,280	950	0.729	57%	56%	3	–
Bahrain	725	12,410	0.846	–	100%	–	–
Bangladesh	140,494	440	0.520	1%	55%	49	68
Barbados	272	9,270	0.878	–	100%	2	–
Belarus	9,832	2,120	0.786	5%	0%	1	–
Belgium	10,405	31,030	0.945	40%	83%	10	–
Belize	283	3,940	0.753	1%	100%	4	–
Benin	6,890	530	0.431	0%	62%	1	1
Bhutan	896	760	0.536	0%	0%	2	5
Bolivia	8,986	960	0.687	0%	0%	15	2
Bosnia and Herzegovina	3,836	2,040	0.786	3%	47%	9	–
Botswana	1,727	4,340	0.565	1%	0%	3	1
Brazil	178,718	3,090	0.792	1%	49%	30	1
Brunei	361	–	0.866	–	100%	–	–
Bulgaria	7,780	2,740	0.808	66%	29%	11	–
Burkina Faso	12,387	360	0.317	3%	0%	2	0
Burma	49,910	–	0.578	0%	49%	4	0
Burundi	7,343	90	0.378	3%	0%	11	–
Cambodia	13,630	320	0.571	0%	24%	8	4
Cameroon	16,400	800	0.497	0%	22%	6	0
Canada	31,902	28,390	0.949	2%	24%	21	0
Cape Verde	481	1,770	0.721	–	100%	–	5
Central African Republic	3,947	310	0.355	0%	0%	5	–
Chad	8,823	260	0.341	0%	0%	2	29
Chile	15,956	4,910	0.854	2%	82%	18	1
China	1,296,500	1,290	0.755	19%	24%	135	2
Colombia	45,300	2,000	0.785	0%	30%	24	1
Comoros	614	530	0.547	–	100%	1	6
Congo	3,855	770	0.512	0%	25%	2	–
Congo, Dem. Rep.	54,775	120	0.385	0%	3%	8	0
Cook Islands	21	–	–	–	–	2	–
Costa Rica	4,061	4,670	0.838	5%	100%	10	2
Côte d'Ivoire	17,142	770	0.420	1%	40%	–	0
Croatia	4,508	6,590	0.841	1%	38%	8	–
Cuba	11,365	–	0.817	14%	100%	13	1
Cyprus	776	17,580	0.891	–	100%	5	–
Czech Republic	10,183	9,150	0.874	19%	0%	5	0
Denmark	5,397	40,650	0.941	20%	100%	2	–
Djibouti	716	1,030	0.495	–	100%	4	18

7 Carbon dioxide emissions				8 Methane emissions			9 Carbon intensity tonnes of CO_2 per \$1,000 GDP		
from burning of fossil fuels million tonnes 2002	tonnes per person 2002	tonnes per person 1950–2000	from transportation million tonnes 2003	million tonnes CO_2e 2000	tonnes CO_2e per person 2000	from agriculture (plus N_2O) million tonnes CO_2e 2003	2003	change 1993–2003	Countries
–	0.0	75	–	13.8	0.5	–	–	–	Afghanistan
4	1.2	195	2	0.5	0.2	0	0.82	-0.45	Albania
73	2.7	1,799	17	28.5	0.9	12.6	1.35	-1.82	Algeria
6	1.2	183	–	14.6	1.2	–	1.36	0.07	Angola
–	7.7	14	–	0.1	1	–	0.75	-0.07	Antigua and Barbuda
115	3.4	4,468	36	83.8	2.3	116.7	0.49	0.00	Argentina
3	2.9	220	1	2.8	0.9	0.9	3.35	-1.66	Armenia
342	19.1	9,188	78	120.8	6.3	103.1	0.91	-0.08	Australia
65	8.9	2,514	21	7.8	1	7.7	0.37	0.01	Austria
26	4.4	1,685	3	11.9	1.5	4.6	5.09	-4.35	Azerbaijan
16	30.9	284	2	2.5	3.8	–	2.37	-0.18	Bahrain
32	0.2	435	4	47.6	0.4	71.5	0.68	0.14	Bangladesh
–	5.9	30	–	0.1	0.4	–	0.62	-0.03	Barbados
55	7.4	3,408	6	12.7	1.3	12.8	6.00	-2.31	Belarus
111	13.7	5,659	26	9.7	0.9	12.3	0.60	-0.06	Belgium
–	3.1	10	–	0.2	0.9	–	0.82	0.30	Belize
2	0.3	20	1	3.1	0.5	–	0.63	0.12	Benin
–	0.2	4	–	1.2	1.4	–	0.69	-0.02	Bhutan
9	1.2	214	3	21.3	2.6	23.3	1.21	0.17	Bolivia
15	3.5	571	–	1.2	0.3	–	–	–	Bosnia and Herzegovina
–	2.1	52	–	6.5	3.9	–	0.66	-0.26	Botswana
309	2.0	7,442	125	298.5	1.8	446	0.55	0.03	Brazil
5	15.0	128	1	2.6	7.8	–	–	–	Brunei
42	6.5	2,793	6	10.2	1.3	5.1	3.58	-0.35	Bulgaria
–	0.1	19	–	8.2	0.7	–	0.36	0.01	Burkina Faso
7	0.2	217	4	61.1	1.3	62.4	0.86	-0.13	Burma
–	0.1	5	–	1.6	0.2	–	0.55	0.14	Burundi
–	0.0	18	–	68	5.4	5	0.13	-0.09	Cambodia
3	0.4	135	2	11	0.7	–	0.59	-0.34	Cameroon
532	19.0	17,424	153	90.5	2.9	60.8	0.80	-0.10	Canada
–	0.3	4	–	–	–	–	–	–	Cape Verde
–	0.1	7	–	6.1	1.6	–	0.36	0.02	Central African Republic
–	0.0	6	–	8.9	1.1	–	0.11	-0.03	Chad
47	3.4	1,218	16	14.5	1	15.2	0.66	0.01	Chile
3,307	2.7	71,766	267	778.4	0.6	1,008.50	2.58	-1.85	China
55	1.2	1,826	18	55.5	1.3	76.8	0.58	-0.18	Colombia
–	0.1	2	–	0.2	0.3	–	–	–	Comoros
1	0.8	48	1	3	0.9	–	0.88	-0.15	Congo
2	0.0	156	0	32.9	0.7	38.7	0.11	-0.12	Congo, Dem. Rep.
–	2.8	1	–	–	–	–	–	–	Cook Islands
5	1.3	104	4	3.6	0.9	–	0.31	-0.03	Costa Rica
5	0.3	134	1	6	0.4	–	0.53	-0.01	Côte d'Ivoire
20	5.0	624	–	3.2	0.7	3.3	1.06	1.06	Croatia
36	3.1	1,157	2	9.1	0.8	7.5	4.85	-0.52	Cuba
6	10.3	141	–	0.3	0.4	–	0.83	-0.10	Cyprus
114	10.9	6,774	17	10.7	1	7.5	1.86	-0.80	Czech Republic
50	10.9	2,498	13	5.7	1.1	10.6	0.36	-0.10	Denmark
–	2.8	10	–	0.6	0.8	–	–	–	Djibouti

95

Countries	1 Population thousands 2004	2 GNI per capita US$ 2004	3 Human Development Index rating 2003	4 Water withdrawn as % of renewable water resources 2002 or latest available data	5 Coastal population as % of total population 1995	6 Weather-related disasters	
						number 2000–05	average annual deaths per million people 1980–2000
Dominica	71	3,650	0.783	–	100%	1	–
Dominican Republic	8,861	2,080	0.749	40%	100%	10	3
East Timor	925	550	0.513	–	–	3	–
Ecuador	13,213	2,180	0.759	4%	61%	9	3
Egypt	68,738	1,310	0.659	96%	53%	4	0
El Salvador	6,658	2,350	0.722	4%	99%	10	9
Equatorial Guinea	506	–	0.655	0%	72%	–	–
Eritrea	4,477	180	0.444	6%	74%	6	–
Estonia	1,345	7,010	0.853	2%	86%	1	–
Ethiopia	69,961	110	0.367	2%	1%	18	287
Fiji	848	2,690	0.752	0%	100%	6	10
Finland	5,215	32,790	0.941	2%	73%	1	–
France	59,991	30,090	0.938	17%	40%	29	0
Gabon	1,374	3,940	0.635	0%	63%	–	–
Gambia	1,449	290	0.470	0%	91%	5	2
Georgia	4,521	1,040	0.732	5%	39%	6	1
Germany	82,631	30,120	0.930	26%	15%	13	–
Ghana	21,053	380	0.520	1%	43%	3	1
Greece	11,075	16,610	0.912	12%	99%	16	0
Grenada	106	3,760	0.787	–	100%	2	–
Guam	171	15,000	–	–	–	3	–
Guatemala	12,628	2,130	0.663	1%	61%	16	6
Guinea	8,073	460	0.466	0%	41%	3	–
Guinea-Bissau	1,533	160	0.348	0%	95%	3	–
Guyana	772	990	0.720	1%	77%	2	–
Haiti	8,592	390	0.475	8%	100%	21	13
Honduras	7,141	1,030	0.667	2%	66%	15	146
Hungary	10,072	8,270	0.862	6%	0%	11	–
Iceland	290	38,620	0.956	0%	100%	–	–
India	1,079,721	620	0.602	36%	26%	85	3
Indonesia	217,588	1,140	0.697	3%	96%	51	1
Iran	66,928	2,300	0.736	54%	24%	24	2
Iraq	25,261	–	–	39%	6%	3	–
Ireland	4,019	34,280	0.946	2%	100%	4	–
Israel	6,798	17,380	0.915	94%	97%	2	–
Italy	57,573	26,120	0.934	22%	79%	15	0
Jamaica	2,665	2,900	0.738	10%	100%	10	3
Japan	127,764	37,180	0.943	21%	96%	31	1
Jordan	5,440	2,140	0.753	128%	29%	4	–
Kazakhstan	14,958	2,260	0.761	31%	4%	5	–
Kenya	32,447	460	0.474	7%	8%	17	1
Kiribati	98	970	–	–	99%	–	–
Korea (North)	22,745	–	–	18%	93%	10	581
Korea (South)	48,142	13,980	0.901	34%	100%	22	3
Kuwait	2,460	17,970	0.844	2450%	100%	–	–
Kyrgyzstan	5,099	400	0.702	22%	0%	7	–
Laos	5,792	390	0.545	0%	6%	4	1
Latvia	2,303	5,460	0.836	1%	75%	3	–

7 Carbon dioxide emissions				8 Methane emissions			9 Carbon intensity		
from burning of fossil fuels million tonnes 2002	tonnes per person 2002	tonnes per person 1950–2000	from transportation million tonnes 2003	million tonnes CO$_2$e 2000	tonnes CO$_2$e per person 2000	from agriculture (plus N$_2$O) million tonnes CO$_2$e 2003	tonnes of CO$_2$ per $1,000 GDP 2003	change 1993–2003	Countries
–	1.4	2	–	0	0.6	–	0.43	0.16	Dominica
18	2.2	321	5	5.9	0.7	–	0.91	0.17	Dominican Republic
–	–	–	–	–	–	–	–	–	East Timor
19	1.8	446	11	16.2	1.3	15.1	1.32	0.17	Ecuador
127	1.9	2,460	32	34.3	0.5	24.7	1.33	-0.15	Egypt
5	0.9	110	3	3.2	0.5	–	0.42	0.10	El Salvador
–	8.0	14	–	0.3	0.7	–	1.60	-4.32	Equatorial Guinea
1	0.2	6	–	–	–	–	–	–	Eritrea
14	13.9	872	2	6.8	5	0.8	2.80	-1.29	Estonia
4	0.1	73	2	43.1	0.7	48	0.61	-0.11	Ethiopia
–	1.7	29	–	1	1.2	–	0.80	0.17	Fiji
63	10.4	2,003	13	5.4	1	5.6	0.43	-0.16	Finland
377	6.8	18,650	139	64.9	1.1	100.2	0.30	-0.03	France
1	3.6	130	0	3.5	2.8	–	0.92	-0.23	Gabon
–	0.2	5	–	0.6	0.5	–	–	–	Gambia
3	0.7	499	1	4.4	0.9	2.3	1.06	-2.04	Georgia
828	10.2	47,314	162	86.5	1.1	89.4	0.45	-0.09	Germany
7	0.3	128	3	6.6	0.3	13.3	–	–	Ghana
90	9.5	2,089	21	11	1	–	0.82	-0.07	Greece
–	3.0	3	–	0	0.3	–	0.59	0.12	Grenada
–	17.7	–	–	–	–	–	0.00	0.00	Guam
10	0.8	169	5	6.2	0.5	–	0.51	0.18	Guatemala
–	0.2	41	–	5.3	0.7	–	–	–	Guinea
–	0.3	5	–	0.8	0.6	–	1.78	-0.08	Guinea-Bissau
–	2.3	60	–	1.4	1.9	–	–	–	Guyana
2	0.2	31	1	3.4	0.4	–	0.48	0.30	Haiti
5	0.9	90	2	5	0.8	–	0.92	0.38	Honduras
55	5.9	3,018	11	9.9	1	7.6	1.13	-0.46	Hungary
2	10.4	81	1	0.3	1	0.3	0.34	-0.04	Iceland
1,016	1.0	18,772	95	428.6	0.4	639.3	1.86	-0.38	India
303	1.4	4,549	70	169.2	0.8	123	1.89	0.30	Indonesia
345	5.4	5,965	88	96.9	1.5	60.8	4.39	0.08	Iran
81	2.7	1,764	21	14.7	0.6	8.9	3.88	-2.17	Iraq
42	10.3	1,188	11	12.8	3.4	19.7	0.37	-0.18	Ireland
63	10.6	1,186	10	11.4	1.8	2.4	0.59	0.03	Israel
432	8.1	14,670	118	35.4	0.6	40.3	0.42	-0.02	Italy
10	4.3	270	2	1.3	0.5	–	1.35	0.24	Jamaica
1,206	9.4	37,265	250	20.7	0.2	34.1	0.25	0.00	Japan
15	3.0	270	4	7.9	1.6	0.5	1.70	0.01	Jordan
141	9.7	8,546	7	27.3	1.8	16.2	6.01	-4.25	Kazakhstan
8	0.3	247	4	19.9	0.7	–	0.81	0.06	Kenya
–	0.3	1	–	0	0.2	–	–	–	Kiribati
68	3.1	3,542	2	34.3	1.5	8.8	–	–	Korea (North)
441	9.8	6,932	98	24.7	0.5	13.9	0.80	-0.16	Korea (South)
57	23.2	1,419	7	9.9	4.5	0	1.43	0.51	Kuwait
5	1.0	364	1	2.2	0.4	0.1	3.25	-4.24	Kyrgyzstan
–	0.2	13	3	6.2	1.2	5	0.49	0.25	Laos
7	3.3	488	4	2.1	0.9	1.5	0.87	-0.62	Latvia

Countries	1 Population thousands 2004	2 GNI per capita US$ 2004	3 Human Development Index rating 2003	4 Water withdrawn as % of renewable water resources 2002 or latest available data	5 Coastal population as % of total population 1995	6 Weather-related disasters number 2000–05	6 Weather-related disasters average annua deaths per million peopl 1980–2000
Lebanon	4,554	4,980	0.759	27%	100%	2	–
Lesotho	1,809	740	0.497	2%	0%	3	1
Liberia	3,449	110	–	0%	58%	–	–
Libya	5,674	4,450	0.799	866%	79%	–	–
Lithuania	3,439	5,740	0.852	1%	23%	2	–
Luxembourg	450	56,230	0.949	40%	0%	–	–
Macedonia	2,062	2,350	0.797	30%	14%	9	–
Madagascar	17,332	300	0.499	5%	55%	17	5
Malawi	11,182	170	0.404	5%	0%	11	2
Malaysia	25,209	4,650	0.796	2%	98%	19	1
Maldives	300	2,510	0.745	–	81%	–	–
Mali	11,937	360	0.333	1%	0%	7	0
Malta	401	12,250	0.867	–	100%	–	–
Marshall Islands	60	2,370	–	–	53%	–	–
Mauritania	2,906	420	0.477	14%	40%	7	58
Mauritius	1,234	4,640	0.791	–	100%	1	–
Mexico	103,795	6,770	0.814	17%	29%	37	2
Micronesia, Fed. Sts.	127	1,990	–	–	98%	4	–
Moldova	4,218	710	0.671	26%	9%	4	1
Mongolia	2,515	590	0.679	1%	0%	9	–
Morocco	30,586	1,520	0.631	40%	65%	8	1
Mozambique	19,129	250	0.379	0%	59%	19	361
Namibia	2,033	2,370	0.627	0%	5%	7	–
Nauru	13	5,000	–	–	–	–	–
Nepal	25,190	260	0.526	14%	0%	9	11
Netherlands	16,250	31,700	0.943	9%	93%	5	–
New Caledonia	219	15,000	–	–	100%	1	–
New Zealand	4,061	20,310	0.933	1%	100%	9	–
Nicaragua	5,604	790	0.690	1%	72%	13	38
Niger	12,095	230	0.281	2%	0%	7	0
Nigeria	139,823	390	0.453	1%	26%	21	0
Niue	2	3,600	–	–	–	1	–
Norway	4,582	52,030	0.963	1%	95%	3	–
Oman	2,659	7,890	0.781	153%	89%	2	1
Pakistan	152,061	600	0.527	70%	9%	30	2
Palau	20	6,870	–	–	–	–	–
Panama	3,028	4,450	0.804	1%	100%	9	–
Papua New Guinea	5,625	580	0.523	0%	61%	5	2
Paraguay	5,782	1,170	0.755	0%	0%	7	1
Peru	27,547	2,360	0.762	1%	57%	18	5
Philippines	82,987	1,170	0.758	12%	100%	45	16
Poland	38,160	6,090	0.858	20%	14%	11	0
Portugal	10,436	14,350	0.904	10%	93%	10	0
Puerto Rico	3,929	18,500	–	–	100%	6	7
Qatar	637	–	0.849	–	100%	–	–
Romania	21,858	2,920	0.792	12%	6%	28	0
Russia	142,814	3,410	0.795	2%	15%	58	0
Rwanda	8,412	220	0.450	15%	0%	6	0

7 Carbon dioxide emissions				8 Methane emissions			9 Carbon intensity		
from burning of fossil fuels million tonnes 2002	tonnes per person 2002	tonnes per person 1950–2000	from transportation million tonnes 2003	million tonnes CO$_2$e 2000	tonnes CO$_2$e per person 2000	from agriculture (plus N$_2$O) million tonnes CO$_2$e 2003	tonnes of CO$_2$ per \$1,000 GDP 2003	change 1993–2003	Countries
15	4.3	336	–	1.3	0.3	–	0.92	0.12	Lebanon
–	0.1	3	–	1.2	0.7	–	0.22	-0.01	Lesotho
–	0.2	36	–	1.1	0.3	–	–	–	Liberia
43	9.0	982	–	8.9	1.7	–	1.47	-0.14	Libya
12	5.5	748	4	3.2	0.9	1.4	1.34	-0.44	Lithuania
9	24.3	592	6	0.5	1.2	0.5	0.53	-0.36	Luxembourg
8	4.1	344	–	1.3	0.6	0	2.36	-1.29	Macedonia
–	0.1	47	–	17.5	1.1	–	0.57	0.22	Madagascar
–	0.1	28	–	3.4	0.3	–	0.48	-0.02	Malawi
116	5.8	1,624	37	31.8	1.4	–	1.43	0.10	Malaysia
–	1.9	4	–	0.1	0.3	–	0.90	0.53	Maldives
–	0.0	15	–	11.1	1	–	0.19	-0.14	Mali
3	7.5	56	1	0.1	0.2	–	0.82	-0.26	Malta
–	–	–	–	–	–	–	–	–	Marshall Islands
–	1.1	54	–	4.1	1.5	–	–	–	Mauritania
–	3.0	40	–	0.2	0.2	–	0.73	0.00	Mauritius
365	3.9	9,398	113	111.7	1.1	50.2	0.69	0.00	Mexico
–	–	–	–	–	–	–	–	–	Micronesia, Fed. Sts.
7	2.5	632	1	2.6	0.6	2.1	6.88	-0.05	Moldova
–	3.2	251	–	8.2	3.4	18.8	7.91	-6.57	Mongolia
33	1.1	655	2	9.3	0.3	–	0.88	-0.04	Morocco
1	0.1	94	1	10.3	0.6	–	0.34	-0.15	Mozambique
2	1.2	18	2	4.2	2.2	–	0.63	0.11	Namibia
–	13.3	4	–	0	0.8	–	–	–	Nauru
3	0.1	34	1	17.1	0.7	24.3	0.50	0.23	Nepal
177	16.1	6,365	34	20.3	1.3	15.8	0.70	-0.09	Netherlands
–	9.0	–	–	–	–	–	–	–	New Caledonia
32	9.9	928	14	26.9	7	35.6	0.67	-0.04	New Zealand
4	0.7	83	2	5.3	1	–	0.96	0.16	Nicaragua
–	0.1	26	–	6	0.6	–	0.64	-0.23	Niger
50	0.8	1,802	25	72.5	0.6	59.9	1.69	-0.70	Nigeria
–	1.4	0	–	–	–	–	–	–	Niue
33	9.9	1,218	13	7	1.6	5	0.26	-0.01	Norway
26	8.6	271	3	3.7	1.5	–	1.11	0.16	Oman
100	0.7	1,848	26	94.7	0.7	148.8	1.34	-0.05	Pakistan
–	–	7	–	0	0.1	–	–	–	Palau
5	4.1	142	3	3.3	1.2	–	1.03	-0.38	Panama
–	0.5	66	–	4	0.8	–	0.68	0.00	Papua New Guinea
4	0.6	68	3	12.3	2.3	–	0.45	0.07	Paraguay
26	1.0	867	9	19.6	0.8	34.4	0.47	-0.10	Peru
70	0.9	1,522	25	36.3	0.5	45.1	0.83	-0.03	Philippines
281	7.4	15,887	29	53.8	1.4	26.9	1.61	-1.64	Poland
63	6.2	1,258	20	8.2	0.8	8.1	0.58	0.02	Portugal
–	9.8	–	–	–	–	–	0.60	0.05	Puerto Rico
28	45.7	531	5	4.2	7.2	–	1.40	-0.97	Qatar
90	4.5	5,882	13	25.7	1.1	11.6	2.33	-1.14	Romania
1,488	11.2	77,120	194	312.3	2.1	97.6	5.24	-0.98	Russia
–	0.1	12	–	2	0.3	–	0.41	0.00	Rwanda

Countries	1 Population thousands 2004	2 GNI per capita US$ 2004	3 Human Development Index rating 2003	4 Water withdrawn as % of renewable water resources 2002 or latest available data	5 Coastal population as % of total population 1995	6 Weather-related disasters number 2000–05	6 Weather-related disasters average annual deaths per million people 1980–2000
Samoa	179	1,860	0.776	–	100%	3	–
São Tomé and Principe	161	370	0.604	–	100%	–	–
Saudi Arabia	23,215	10,430	0.772	814%	30%	6	–
Senegal	10,455	670	0.458	4%	83%	6	–
Serbia and Montenegro	8,152	2,620	–	7%	8%	7	–
Seychelles	85	8,090	0.821	–	100%	1	–
Sierra Leone	5,436	200	0.298	0%	55%	2	–
Singapore	4,335	24,220	0.907	–	100%	–	–
Slovakia	5,390	6,480	0.849	2%	0%	6	0
Slovenia	1,995	14,810	0.904	7%	61%	–	–
Solomon Islands	471	550	0.594	–	100%	3	17
Somalia	9,938	–	–	5%	55%	15	20
South Africa	45,584	3,630	0.658	27%	39%	20	1
Spain	41,286	21,210	0.928	32%	68%	13	0
Sri Lanka	19,444	1,010	0.751	20%	100%	11	2
St. Kitts and Nevis	47	7,600	0.834	–	–	–	–
St. Lucia	164	4,310	–	–	100%	1	22
St. Vincent and the Grenadines	108	3,650	0.755	–	100%	3	–
Sudan	34,356	530	0.512	12%	3%	7	295
Suriname	443	2,250	0.755	0%	87%	–	–
Swaziland	1,120	1,660	0.498	–	31%	3	4
Sweden	8,985	35,770	0.949	2%	88%	3	–
Switzerland	7,382	48,230	0.947	2%	0%	8	–
Syria	17,783	1,190	0.721	27%	35%	4	–
Tajikistan	6,430	280	0.652	15%	0%	16	–
Tanzania	36,571	330	0.418	1%	21%	9	1
Thailand	62,387	2,540	0.778	8%	39%	33	2
Togo	4,966	380	0.512	1%	45%	–	–
Tonga	102	1,830	0.810	–	98%	2	–
Trinidad and Tobago	1,323	8,580	0.801	8%	100%	3	–
Tunisia	10,012	2,630	0.753	61%	84%	2	1
Turkey	71,727	3,750	0.750	15%	58%	24	0
Turkmenistan	4,931	1,340	0.738	39%	8%	–	–
Tuvalu	11	1,100	–	–	–	–	–
Uganda	25,920	270	0.508	0%	0%	16	1
Ukraine	48,008	1,260	0.766	19%	21%	7	0
United Arab Emirates	4,284	–	0.849	1669%	85%	–	–
United Kingdom	59,405	33,940	0.939	8%	99%	17	–
United States	293,507	41,400	0.944	17%	43%	146	1
Uruguay	3,399	3,950	0.840	1%	79%	10	–
Uzbekistan	25,930	460	0.694	51%	3%	3	–
Vanuatu	215	1,340	0.659	–	100%	4	30
Venezuela	26,127	4,020	0.772	1%	73%	10	69
Vietnam	82,162	550	0.704	6%	83%	43	8
Virgin Islands (US)	113	17,200	–	–	100%	1	–
Yemen	19,763	570	0.489	71%	64%	10	4
Zambia	10,547	450	0.394	1%	0%	8	0
Zimbabwe	13,151	–	0.505	9%	0%	7	0

7 Carbon dioxide emissions				8 Methane emissions			9 Carbon intensity		
from burning of fossil fuels million tonnes 2002	tonnes per person 2002	tonnes per person 1950–2000	from transportation million tonnes 2003	million tonnes CO_2e 2000	tonnes CO_2e per person 2000	from agriculture (plus N_2O) million tonnes CO_2e 2003	tonnes of CO_2 per $1,000 GDP 2003	change 1993–2003	Countries
–	0.8	3	–	–	–	–	–	–	Samoa
–	0.6	2	–	–	–	–	–	–	São Tomé and Principe
301	13.5	4,625	63	54.4	2.6	10	1.60	0.22	Saudi Arabia
4	0.5	87	1	8.4	0.9	11.6	0.89	0.00	Senegal
49	4.8	1,887	–	8.1	0.8	–	1.41	0.00	Serbia and Montenegro
–	14.6	5	–	0	0.3	–	1.88	0.89	Seychelles
–	0.2	23	–	2.4	0.5	–	0.92	-0.04	Sierra Leone
42	27.9	949	6	1.2	0.3	0.1	1.26	-0.07	Singapore
38	7.1	2,311	6	4.5	0.8	4.1	1.67	-1.20	Slovakia
15	8.4	501	–	2.3	1.2	2.1	0.80	-0.13	Slovenia
–	0.4	4	–	0.2	0.4	–	0.60	0.60	Solomon Islands
–	0.1	–	–	–	–	–	–	–	Somalia
301	9.1	10,202	40	36.8	0.8	39.8	2.97	-0.03	South Africa
303	8.3	7,685	103	39.3	1	43.6	0.56	0.05	Spain
11	0.6	203	5	13.9	0.8	–	0.65	0.11	Sri Lanka
–	2.6	2	–	–	–	–	0.30	-0.05	St. Kitts and Nevis
–	2.4	5	–	0.1	0.3	–	0.52	0.23	St. Lucia
–	1.5	2	–	0	0.4	–	0.49	0.21	St. Vincent and the Grenadines
8	0.3	167	4	43.1	1.4	–	–	–	Sudan
–	3.9	73	–	0.9	2	–	2.14	-0.50	Suriname
–	1.3	14	–	1	1	–	0.92	0.40	Swaziland
50	6.3	3,025	22	5.9	0.7	8.9	0.22	-0.05	Sweden
40	6.0	1,757	16	4.4	0.6	5.5	0.17	-0.02	Switzerland
47	2.9	888	12	9.7	0.6	–	2.21	-0.14	Syria
5	1.0	450	3	1.4	0.2	0.2	4.93	1.02	Tajikistan
3	0.1	91	2	29.3	0.9	–	0.33	-0.06	Tanzania
179	3.1	2,435	51	75.9	1.3	63.8	1.39	0.26	Thailand
1	0.2	21	1	2	0.4	–	0.88	0.30	Togo
–	1.2	2	–	0.1	1	–	–	–	Tonga
17	24.9	469	2	3.1	2.4	–	3.61	3.61	Trinidad and Tobago
19	2.2	399	4	4.4	0.5	–	0.97	-0.10	Tunisia
193	2.9	4,090	36	90.7	1.3	69.2	0.98	0.09	Turkey
40	8.8	1,455	2	27.1	5.8	2.3	5.45	2.30	Turkmenistan
–	–	–	–	–	–	–	–	–	Tuvalu
–	0.1	40	–	12.4	0.5	17.1	0.22	-0.06	Uganda
292	7.1	20,768	18	148	3	36	8.85	-1.33	Ukraine
89	44.1	1,171	15	35.2	10.8	0	1.69	-0.03	United Arab Emirates
528	9.5	29,843	133	48.8	0.8	49.4	0.37	-0.13	United Kingdom
5,635	20.0	212,826	1,794	614.4	2.2	469.9	0.56	-0.13	United States
4	1.6	256	2	18.3	5.5	17.7	0.31	0.03	Uruguay
118	4.4	5,017	11	46.2	1.9	20.2	7.35	-2.14	Uzbekistan
–	0.4	2	–	0.5	2.3	–	0.39	0.06	Vanuatu
123	5.5	4,264	36	91.7	3.8	34.3	1.42	0.41	Venezuela
57	0.8	888	16	69.9	0.9	60.4	1.60	0.17	Vietnam
–	121.5	–	–	–	–	–	–	–	Virgin Islands (US)
11	0.5	252	5	8.7	0.5	–	0.93	-0.66	Yemen
2	0.2	170	1	10.3	1	–	–	–	Zambia
11	0.9	476	–	10.2	0.8	–	0.98	-1.32	Zimbabwe

101

Sources

DEFINITION OF KEY TERMS
The glossary published by the Intergovernmental Panel on Climate Change is the principal source for the technical definitions: www.ipcc.ch/pub/gloss.pdf and www.ipcc.ch/pub/syrgloss.pdf.

Other sources for technical terms include: www.ametsoc.org/amsedu/WES/glossary.html, amsglossary. allenpress.com/glossary, cdm.unfccc.int/, www.climatechange.ca.gov/glossary, www.evomarkets.com/ghg_glossary.html, ilrdss.sws.uiuc.edu/glossary/glossary_allresults.asp, www.ngdc.noaa.gov/seg/hazard/stratoguide/glossary.html, www.vulnerabilitynet.org.

Regional groups are defined in: www.cisstat.com/eng, europa.eu.int/abc/governments/index_en.htm, www.un.org/special-rep/ohrlls/ldc/list.htm, www.oecd.org, www.opec.org. The UNFCCC web site is also useful: www.unfccc.int

Part 1: Signs of Change

Quote: http://www.environmentaldefense.org

20–21 WARNING SIGNS
International Panel on Climate Change (IPCC) Fourth Assessment Report, Working Group I, *The Physical Science Basis*. Summary for Policymakers, IPCC, 2007 www.ipcc.ch

International Panel on Climate Change (IPCC) Fourth Assessment Report, Working Group II, *Climate Change 2007: Impacts, Adaptation and Vulnerability*. Summary for Policymakers, IPCC, 2007 www.ipcc.ch

International Panel on Climate Change (IPCC) Fourth Assessment Report, Working Group III, *Mitigation*. Summary for Policymakers. IPCC, 2007. www.ipcc.ch

Alaskan permafrost
Arctic Climate Impact Assessment (ACIA), *Impacts of a Warming Arctic*, 2004, Cambridge University Press. Reuters, Warming climate disrupts Alaska natives' lives, quoting Gunter Weller, director of the University of Alaska Fairbanks' Center for Global Change and Arctic System Research, 2004: www.planetark.com/dailynewsstory.cfm/newsid/24761/story.htm

Atlantic hurricanes
National Oceanographic and Atmosphere Administration, US Department of Commerce:
www.noaanews.noaa.gov/stories2005/s2540.htm

Asian summer monsoon
World Meteorological Organization (WMO), statement on the status of global climate, 2004.

Canadian polar bears
World Meteorological Organization (WMO) *Statement on the Status of the Global Climate in 2006*. WMO-No. 1016. Geneva, 2007. www.wmo.ch/news/download/Status_2006_E.pdf

Changes in China
Y Shi et al, Recent and future climate change in northwest China, *Climatic Change*, 2007, 80: 379–393.

Coral bleaching
C Wilkinson et al, International Society for Reef Studies (ISRS), Ecological and socioeconomic impacts of 1998 coral mortality in the Indian Ocean: An ENSO impact and a warning of future change? *Ambio*, 1999, 28: 188–96.
ISRS, statement on global coral bleaching in 1997–98: www.uncwil.edu/isrs/

Drought in Australia
http://www.guardian.co.uk/australia/story/0,,2061761,00.html#article_continue

Drought in China
WMO 2007, op. cit.

Drought in Southern Brazil
WMO 2007, op. cit.

Drought in Horn of Africa
WMO 2007, op.cit.

European Alps
World Meteorological Organization (WMO), statement on the status of global climate, 2003.
www.wmo.ch/web/wcp/wcdmp/statement/html/966_E.pdf

European autumn
WMO 2007, op cit.

European butterfly ranges
C Parmesan et al, Poleward shifts in geographical ranges of butterfly species associated with regional warming, *Nature*, 1999, 399: 579–83.

European heatwave
www.ifrc.org/publicat/wdr2004/chapter2.asp

Floods in Bolivia
WMO 2007, op. cit.

Floods in Brazil
World Meteorological Organization (WMO), Statement on the status of global climate, 2004. www.wmo.ch/web/wcp/wcdmp/statement/html/983_E.pdf Accessed May 2006

Floods in China
www.reliefweb.int/rw/RWB.NSF/db900SID/VBOL-6DKJAS?OpenDocument

Floods in Horn of Africa
WMO 2007, op. cit.

Floods in New Zealand
twm.co.nz/wet2004.html#Flood

Indian heatwave
World Meteorological Organization (WMO), statement on the status of global climate, 2004.

Larsen B ice shelf
nsidc.org/iceshelves/larsenb2002
E Rignot, Changes in ice dynamics and mass balance of the Antarctic ice sheet. *Philosophical Transactions of the Royal Society a-Mathematical Physical and Engineering Sciences* 364, 2006, 1637–55.

Mosquitoes
W E Bradshaw, C M Holzapfel, Genetic shift in photoperiodic response correlated to global warming, 2001, *Proceedings of the National Academy of Sciences of the United States of America* (*PNAS*): www.pnas.org, 10.1073/pnas.241391498. www.nature.com/nsu/nsu_pf/011108/011108-6.html)

Siberian melt
S N Kirpotin et al, Sub-Arctic palsas as indicator of climatic changes, INTAS report 2005 from Tomsk State University, Tomsk, Russia; Institute of Soil Science and Agrochemistry of the Siberian branch of Russian Academy of Science, Novosibirsk, Russia; Yugorskiy State University, Khanty–Mansiysk, Russia; University of Utrecht, Utrecht, Netherlands.

South Atlantic hurricane
www.metoffice.com/sec2/sec2cyclone/catarina.html

Spanish drought
www.wmo.ch/index-en.html, accessed 26 May 2005.

Tropical Andes
Intergovernmental Panel on Climate Change (IPCC), Climate Change 2001: The Scientific Basis www.grida.no/climate/ipcc_tar/wg1/064.htm.

Washington, DC flowering
Mones S A et al, Earlier plant flowering in spring as a response to global warming in the Washington, DC area, 2001, *Biodiversity and Conservation* 10: 597-612.

22-23 POLAR CHANGES
Arctic Climate Impact Assessment: www.acia.uaf.edu/
Lloyd Peck, British Antarctic Survey, quoted in the *Guardian*, 19 October 2005: 9.
C Krauss et al, As Polar Ice Turns to Water, Dreams of Treasure Abound. *New York Times*, October 10, 2005
D Massey , 'Ice is melting everywhere', *People and Planet*, 3 May 2005. http://www.peopleandplanet.net/doc.php?id=2470

Antarctic
Lloyd Peck, *Guardian* op cit.
M P Meredith, J C King, Rapid climate change in the ocean to the west of the Antarctic Peninsula during the second half of the 20th century, *Geophysical Research Letters*, 2005, 32 (19), L19604, 10.1029/2005GL02042: 5.
D G Vaughan, J R Spouge, Risk estimation of collapse of the West Antarctic ice sheet, *Climate Change*, 2002, 52: 65-91.
Goddard Institute for Space Studies, NASA Goddard Space Flight Center, Earth Sciences Directorate, Global Temperature Anomalies in .01 C <http://www.giss.nasa.gov/data>, updated January 2005.

Arctic
NASA: www.nasa.gov/centers/goddard/news/topstory/2003/1023esuice.html www.ngo.grida.no/wwfap/polarbears/risk/climate.html
Arctic Climate Impact Assessment: www.acia.uaf.edu

"As long as it's ice..."
C Krauss et al, As Polar Ice Turns to Water, Dreams of Treasure Abound, *New York Times*, October 10, 2005. Accessible via: www.solarliving.org/enews/56.htm

Greenland ice cap
K Steffen, R Huff, Greenland melt extent, 2005, Cooperative Institute for Research in Environmental Sciences (CIRES), University of Colorado at Boulder: cires.colorado.edu/science/groups/steffen/greenland/melt2005
K Steffen et al, The melt anomaly of 2002 on the Greenland Ice Sheet from active and passive microwave satellite observations, *Geophysical Research Letters*, 2004, 31 (20), L20402, 10.1029/2004GL020444.
H Hanna et al, Runoff and mass balance of the Greenland ice sheet, *Journal of Geophysical Research 1958-2003*, 2005, 110, D13108, 10.1029/2004JD005641.

Area of Greenland melt
Steffen Research Group: cires.colorado.edu/science/groups/steffen/greenland/melt2005

24-25 GLACIAL RETREAT
M B Dyurgerov, M F Meier, Twentieth century climate change: evidence from small glaciers, *Proceedings of the National Academy of Sciences of the United States of America* (*PNAS*), 2000, 97: 1406-411.
N G Glasser et al, Photographic evidence of the return period of the Svalbard surge–type glacier: a tributary of Pedersenbreen, Kongsfjord. *Journal of Glaciology*, 2005, 50: 169.
C Harris et al, Warming permafrost in European mountains, *Global and Planetary Change*, 2003, 39: 215-25.
S Harrison et al, Onset of rapid calving and retreat of Glacier San Quintin, Hielo Patagónico Norte, southern Chile, *Polar Geography*, 2001, 25 (1): 54-61.

E Rignot, A Rivera, G Cassassa, Contribution of the Patagonia Icefields to sea level rise, *Science*, 2003, 302: 434-37.
H Schroeder, I Severskiy, eds. *Water resources in the basin of the Ili River (Republic of Kazakhstan)*, 2004, Mensch and Buch Verlag, Berlin.

Glacial change
World Global Monitoring Service, *Fluctuations of Glaciers (FoG) 1995-2000*: www.geo.unizh.ch/wgms/fog.html
WWF, Going, Going, Gone. Climate Change and Global Glacier Decline: www.panda.org
L Mastny, Worldwatch Institute, Melting of Earth's ice cover reaches new high: www.upe.ac.za/botany/pssa/ice_melt.htm

USA
earthobservatory.nasa.gov/Newsroom/NewImages/images.php3?img_id=16970

Images from:
R S Brackett, Arapaho Glacier, 1898, published in R S Waldrop, *Arapaho Glacier: A Sixty Year Record*, 1964, University of Colorado Studies, Series in Geology (3).
T Pfeffer, Arapaho Glacier, 2003.

Glacial lakes
www.visibleearth.nasa.gov

26-27 EVERYDAY EXTREMES
S Curtis, R Adler, The magnitude and variability of global and regional precipitation based on the 21 year Global Precipitation Climatology Project (GPCP) and 3 year Tropical Rainfall Measuring Mission (TRMM) data sets, 2001, Annual Meeting of the American Meteorological Society, Albuquerque, NM.
EM-DAT: OFDA/CRED international disaster database, Université Catholique de Louvain, Brussels: www.em-dat.net
M Beniston, Department of Geosciences, University of Fribourg, Switzerland, and D B Stephenson, Department of Meteorology, University of Reading, UK, Extreme climatic events and their evolution under changing climatic conditions, 2004.
C Parmesan and H Galbraith, Observed impacts of global climate change in the US, Pew Center on Global Climate Change, 2004.

Disasters
Deaths
Floods and windstorms
EM-DAT op cit. Note that the recording of climatic disasters has improved over time. The US Office of Foreign Disaster Assistance and the CRED began sharing their information in 1973 and the EM-DAT was created as a joint monitoring service in 1988. The rules for recording multi-country disasters were changed in 2003. At the same time, the global population and the number of people in hazard prone regions have both increased. Thus, the rising trend in disasters is likely to be a result of more frequent and severe geophysical events, more people exposed, and better monitoring and recording of disasters. See: W N Adger and N Brooks, Does global environmental change cause vulnerability to disaster? In M Pelling (ed.) Natural Disaster and Development in a Globalising World, 2003, London, Routledge, pp 19-42. www.cru.uea.ac.uk/~e118/publications/Pelling-adger-brooks-draft.pdf

2003 heatwave in Western Europe
Image by R Stöckli, R Simmon D Herring, NASA Earth Observatory, based on data from the MODIS land team.
EM-DAT op cit.
www.ifrc.org/publicat/wdr2004/chapter2.asp

Part 2: Forcing Change
Quote: International Panel on Climate Change (IPCC) Fourth Assessment Report, Working Group I, The Physical Science Basis. Summary for Policymakers, IPCC, 2007 www.ipcc.ch

30–31 THE GREENHOUSE EFFECT
J Houghton, *Global Warming: The Complete Briefing*, 2004, Cambridge University Press, 3rd edn.

Siberian Peat Proves Key Player for Greenhouse Gas Production in Past, Future, UCLA-Led Team Finds: www.newsroom.ucla.edu/page.asp?RelNum=4865

Change in atmospheric composition since the Industrial Revolution
Atmospheric lifetime
Contribution of different gases to radiative forcing since the Industrial Revolution
cdiac.esd.ornl.gov/pns/current_ghg.html

32–33 THE CLIMATE SYSTEM
ACIA, Artic Climate Impact Assessment. The Impacts of a Warming Arctic: Artic Climate Impact Assessment. Cambridge, Cambridge University Press, 2004.

R B Alley et al, Abrupt climate change, *Science, 2002,* 299 (5615): 2005–2010.

W S Broecker, Thermohaline circulation, the Achilles heel of our climate system: will man-made CO_2 upset the current balance? *Science,* 1997, 278: 1582–588.

A V Fedorov, SG Philander, Is El Nino changing? *Science,* 2000, 288: 1997–2002.

T F Stocker, O Marchal, Abrupt climate change in the computer: is it real? *Proceedings of the National Academy of Sciences of the United States of America (PNAS),* 2000, 97: 1362–365.

R T Sutton, D L R Hodson, Atlantic ocean forcing of North American and European summer climate. *Science,* 2005, 309: 115–18.

K Trenberth, Uncertainty in hurricanes and global warming, *Science,* 2005, 380 (5729): 1753–754.

Storms
Trenberth, 2005, op. cit.

Northern Europe
Stocker and Marchal, 2000.

Harry Bryden, National Oceanography Centre, Southampton, cited in the *Guardian,* 1 December 2005: 3.

Sahel, Europe
Sutton and Hodson, 2005.

El Niño
Fedorov and Philander, 2000.

34–35 INTERPRETING PAST CLIMATES
Intergovernmental Panel on Climate Change, Climate Change 2001: Working Group I: *The Scientific Basis, Summary for Policy Makers,* 2001. Cambridge University Press, Cambridge.

T J Osborn, K R Briffa, The spatial extent of the twentieth century warmth in the context of the past 1200 years. *Science,* 2006, 311 (5762): 841–44.

"CO_2 is about 30% higher..."
Thomas Stocker, quoted by Richard Black, BBC News website, November 24, 2005: news.bbc.co.uk/2/hi/science/nature/4467420.stm

CO_2 fluctuations
Link between CO_2 and temperature
Accounting for warming

A millennium of warming in Northern Hemisphere
IPCC: www.ipcc.ch/present/graphics.htm

Accumulated knowledge
Web of Science – Science and Social Science Citation Indices. www.isiknowledge.com

36–37 FORECASTING FUTURE CLIMATES
International Panel on Climate Change (IPCC) Fourth Assessment Report, Working Group I, *The Physical Science Basis.* Summary for Policymakers, IPCC, 2007 www.ipcc.ch

K Ruosteenoja et al, Future climate in world regions: an inter-comparison of model-based projections for the new IPCC emissions scenarios, 2003, Finnish Meteorological Institute, Helsinki: ipcc-ddc.cru.uea.ac.uk/asres/scatter_plots/scatterplots_home.htm

Hadley Centre, Uncertainty, risk and dangerous climate change; recent research on climate change science, 2004. Hadley Centre, Exeter.

D A Stainforth, et al, Letter: Uncertainty in predictions of the climate response to rising levels of greenhouse gases, *Nature* 433, 2005, 403–406 www.nature.com/nature/links/050127/050127-6.html

Nebojsa Nakicenovic and Rob Swart, eds. *Special Report on Emissions Scenarios 1996* IPCC and Cambridge University Press www.grida.no/climate/ipcc/emission

Local warming
D Stainforth, personal communication. www.ClimatePrediction.net

CO_2
Projected warming
IPCC Working Group I, 2007, SPM Figure 5 www.ipcc.ch/present/graphics.htm

Scientific confidence in predictions of climate change
IPCC, 2007, op cit. The IPCC adopted a formal definition of uncertainties: Virtually certain: probability of being true is greater than 99%; Extremely likely: greater than 95%; Very likely: greater than 90%; Likely: greater than 66%.

Precipitation
D Stainforth, personal communication. www.ClimatePrediction.net

Part 3: Driving Climate Change
Quote: www.earthfuture.com/stormyweather/quotes/#a

40–41 EMISSIONS PAST AND PRESENT
Cumulative carbon emissions
World Resources Institute, Washington DC, Climate Analysis Indicators Tool (CAIT) version 2.0: cait.wri.org/, accessed June 2005.

Sources of greenhouse gas emissions
Energy emissions
Comparative growth
World Resources Institute, Washington DC, Climate Analysis Indicators Tool (CAIT) version 3.0: cait.wri.org/, accessed March 2006.

CO_2 in the atmosphere
UNEP Vital Graphics, quoting David J Hofmann of the Office of Atmospheric Research at the National Oceanic and Atmospheric Administration, March 2006: www.grida.no/climate

42–43 FOSSIL FUELS
Fossil fuel burning
CO_2 emitted

CO_2 per person

International Energy Agency:
Approach contains total CO_2 emissions from fuel combustion as calculated using the IPCC Reference Approach, and corresponds to IPCC Source/Sink Category 1A with the following exception. The Reference Approach is based on the supply of energy in a country and as a result, all inventories calculated using this method include fugitive emissions from energy transformation (eg from oil refineries) which are normally included in Category 1B. For this reason, Reference Approach estimates are likely to overestimate national CO_2 emissions.

Coal reserves

Global Coal Reserves. Energy Information Administration, *International Energy Annual 2003,* table posted June 2005. Global coal reserves are defined by the World Energy Council as the tonnage within the Proved Amount in Place that can be recovered (extracted from the earth in raw form) under present and expected local economic conditions with existing available technology. It approximates the US term proved (measured) reserves, coal.

44-45 METHANE AND OTHER GASES

Methane emissions
Methane

World Resources Institute, CAIT version 3.0, 2006, Washington DC. World Resources Institute, Washington DC, Climate Analysis Indicators Tool (CAIT) cait.wri.org, accessed March 2006.

Methane hydrates

N Goel, In situ methane hydrate dissociation with carbon dioxide sequestration: current knowledge and issues, *Journal of Petroleum Science and Engineering*, 2006: www.elsevier.com

G P Glasby, Potential impact on climate of the exploitation of methane hydrate deposits offshore, *Marine and Petroleum Geology*, 2003, 20: 163-75.

H Kono et al, Synthesis of methane gas hydrate in porous sediments and its dissociation by depressurizing, *Powder Technology*, 2002, 122: 239-46.

Nitrous oxide

International Energy Agency, IEA Energy Information Centre, on-line data services, http://data.iea.org/ieastore/default.asp

Global warming potential

Intergovernmental Panel on Climate Change, Climate Change 2001: Working Group I: *Technical Summary,* 2001. Cambridge University Press, Cambridge. p 47.

46-47 TRANSPORTATION

EIA, *Household Transport Report, 2005.* www.eia.doe.gov/emeu/rtecs/nhts_survey/2001/

Transport emissions
Increasing emissions
International shipping and aviation

International Energy Agency, 2006, CO_2 Emissions from Fuel Combustion (edition 2005), Paris, France, http://data.iea.org/stats/eng/main.html

Air travel

Airports Council International, *ACI Passenger and Freight Forecasts 2005-2020*, accessed 23 August 2005.

48-49 DISRUPTING THE CARBON BALANCE

R A Houghton, C L Goodale, Effects of land-use change on the carbon balance of terrestrial ecosystems in R S DeFries, G P Asner, R A Houghton (eds), *Ecosystems and Land Use Change*,

2004: 85-98, American Geophysical Union, Washington DC. FAO, Carbon sequestration in dryland soils: World soil resources reports, 2004: 129

Woods Hole Research Center: www.whrc.org

Carbon cycle
Woods Hole Research Center op cit.

Carbon release
Carbon Exchange
Carbon Dioxide Information Analysis Center (CDIAC): cdiac.esd.ornl.gov/ftp/trends/landuse/houghton/houghtondata.txt

50-51 AGRICULTURE

Agriculture

World Resources Institute, CAIT version 3.0, 2006, Washington DC. World Resources Institute, Washington DC, Climate Analysis Indicators Tool (CAIT) http://cait.wri.org, accessed March 2006.

UNDP, New York, Human Development Report 2005: http://hdr.undp.org/statistics/highlights.

Beef consumption

Food and fuel in a warmer world: www.fao.org/News/2001/011109-e.htm

USDA figure for late 1990s, quoted by WorldWatch Institute: www.worldwatch.org/press/news/1998/07/02/#1

The luxury of choice
CO_2 emissions
P Watkiss, AEA Technology Environment, The validity of food miles as an indicator of sustainable development, July 2005, table E4, Defra: statistics.defra.gov.uk/esg/reports/foodmiles/final.pdf

Part 4: Expected Consequences

Quote: www.iucn.org/en/news/archive/2003/newapril03.htm

54-55 DISRUPTED ECOSYSTEMS

B Walker, W Steffen, An overview of the implications of global change for natural and managed terrestrial ecosystems, *Conservation Ecology*, 1997, 1 (2): 2 and at: www.consecol.org/vol1/iss2/art2

J R Malcolm et al, Habitats at risk: global warming and species loss in globally significant terrestrial ecosystems, 2002, World Wide Fund for Nature (WWF), Gland, Switzerland.

C D Thomas et al, Extinction risk from climate change, *Nature*, 2004, 427: 145-48.

R E Green et al, eds, Global climate change and biodiversity, April 2003, University of East Anglia, Norwich, UK.

"Between 20%..."
International Panel on Climate Change, Working Group II Contribution to the Intergovernmental Panel on Climate Change Fourth Assessment Report, *Climate Change 2007: Impacts, Adaptation and Vulnerability*. Summary for Policymakers, IPCC, 2007 www.ipcc.ch

"At least 40%..."
www.globalissues.org/EnvIssues/Biodiversity/WhoCares.asp quoting from Convention on Biological Diversity: www.biodiv.org/convention/articles.asp

"The loss of each..."
Climate change, www.iucn.org/en/news/archive/2001_2005/mbclimate.pdf

Global warming
Rik Leemans, Bas Eickhout, Another reason for concern: regional and global impacts on ecosystems for different levels of climate change, *Global Environment Change* 14 (2004) 219-228, www.elsevier.com/locate/gloenvcha

Central America
Mounts Kenya and Kilimanjaro
Non-native species
Ocean warming
South Africa
Conservation International, Conservation Frontlines: Current and projected effects of climate change: www.conservation.org/xp/ frontlines/science/08110405.xml
www.biodiversityhotspots.org/xp/Hotspots

Mountains of Central Asia
Polynesia and Micronesia
Brazilian Cerrado
Conservation International, biodiversity hotspots: www. biodiversityhotspots.org/xp/Hotspots/hotspots_by_region

Sundarbans Delta
WWF: powerswitch.panda.org/the_problem/tiger.cfm

Europe
www.birdlife.org/news/news/2004/01/climate_change.html
news.bbc.co.uk/2/hi/science/nature/3375447.stm

56–57 THREATENED WATER SUPPLIES
World Resources Institute, earthtrends, water resources and freshwater ecosystems: earthtrends.wri.org

UNEP, vital water: www.unep.org/vitalwater/21.htm

J Alcamo, T Henrichs, Critical regions: A model-based estimation of world water resources sensitive to global changes, *Water Policy*, 2002, 64: 352–62.

J Alcamo, T Henrichs, T Rösch, World water in 2025: global modeling and scenario analysis for the world commission on water for the 21st century, 2000, Kassel World Water Series Report 2, Center for Environmental Systems Research, University of Kassel, Germany.

N W Arnell, Climate change and global water resources: SRES emissions and socio-economic scenarios, *Global Environmental Change*, 2004, 14: 31–52.

www.unesco.org/water/news/newsletter/123.shtml#news_1, accessed 11 December 2005. (Based on work of Lisa Mastny, Worldwatch Institute, news brief 00–02).

Water use
World Resources Institute (WRI), 2000, compiled 2003 by the AQUASTAT global information system on water and agriculture: www.wri.org

Freshwater resources
World Bank: *World Development Indicators 2005*, supplemented with data from Table: Freshwater Resources 2005, EarthTrends, World Resources Institute, earthtrends.wri.org/

European Environment Agency, Environment in the European Union at the turn of the century, 3.5 Water stress, *Environmental assessment report* 2, 1999.

FAO, Crops and Drops, European Environment Agency, 2002, citing OECD 1998.

River flow
Met Office and Defra. Climate change, rivers and rainfall; Recent research on climate change science from the Hadley Centre, 2005. Brochure prepared for COP11. Model results are for the new Hadley Centre model, HadGEM1.

"Future climate change..."
Quoted on: M Rowling, Climate change to create African "water refugees" March 22, 2006, Reuters Foundation AlertNet: www. alertnet.org/thefacts/reliefresources/114303555233.htm

58–59 FOOD SECURITY
F N Tubiello and G Fisher, Reducing climate change impacts on agriculture: global and regional effects of mitigation 2000–2080, 2006 (draft paper), IIASA.

G Fischer, M Shah, H van Velhuizen, Climate change and agricultural vulnerability, 2002 (working paper) IIASA, Laxenbourg. Based on Hadley Centre model H adCM3

G Fischer et al, Socio-economic and climate change impacts on agriculture: an integrated assessment 1990–2080, *Philosophical Transactions of the Royal Society B: Biological Sciences*, 2005 (published online):
www.pubs.royalsoc.ac.uk/index.cfm?page=1085

Future food production
F N Tubiello and G Fisher op cit, table 8.
Range of potential impacts for two climate scenarios (Hadley, CSIRO) and a reference projection of cereal production (A2) for the 2050s to 2080s, assuming agronomic and economic adaptation to climate change.

Current malnutrition
UNICEF, *Progress for Children: A report card on nutrition*, no. 4, May 2006, pp30–31: www.unicef.org/progressforchildren/2006n4

Africa
G Fischer, M Shah, H van Velhuizen op cit.
Map based on HadCM3 climate model.

60–61 THREATS TO HEALTH
R Akhatar et al, Human health, from J J McCarthy et al, *Climate Change 2001: Impacts, Adaptation, and Vulnerability*, 2001, 9: 451–86, Cambridge University Press, Cambridge.

J Brownstein, Lyme disease: implications of climate change, in P Epstein and E Mills, eds, *Climate Change Futures: Health, Ecological, and Economic Dimensions*, 2005 (Cambridge MA): 45–47, Center for Health and the Global Environment and Harvard Medical School.

K L Ebi et al, Infectious and respiratory diseases: malaria, in P Epstein and E Mills op cit: 32–41.

P Epstein and E Mills, eds, *Climate Change Futures: Health, Ecological, and Economic Dimensions*, 2005 (Cambridge MA): Center for Health and the Global Environment and Harvard Medical School.

S J Kutz et al, Emerging parasitic infections in arctic ungulates, *Integrative and Comparative Biology*, April 2004, 44 (2): 109–18.

World Health Organization (WHO), *World Malaria Report* 2005, WHO, Geneva: rbm.who.int/wmr2005/html/exsummary_en.htm, accessed 19 December 2005.

Health impact of climate change
Total health impact
D H Campbell-Lendrum, C F Corvalán, A Prüss-Ustün, How much disease could climate change cause? in A J McMichael et al, eds, *Climate Change and Human Health*, 2003: 133–58. WHO, Geneva.

Tick-borne diseases
J Brownstein, op cit.

Malaria
WHO, *World Malaria Report*, op cit.
K L Ebi, op cit.

62–63 RISING SEA LEVELS
Projected sea-level rise
IPCC: www.ipcc.ch/present/graphics.htm

Total land loss
Effect of sea-level rise
Dynamic and Interactive Assessment of National, Regional and

Global Vulnerability of Coastal Zones to Climate Change and Sea-Level Rise (DINAS-COAST) project: www.dinas-coast.net, using the global assessment model, DIVA 2.0 (Dynamic and Interactive Vulnerability Assessment)
Richard J.T. Klein, Poul Grashoff, personal communication

Nile Delta
UNEP vital climate graphics: www.grida.no/climate/vital

Evacuated islands
www.jamaica-gleaner.com/gleaner/20060127/news/news1.html

Disappearing islands
news.bbc.co.uk/2/hi/science/nature/368892.stm

Abandoned island
www.sidsnet.org/1f.html

Threatened island
www.tuvaluislands.com/news/archives/2006/2006-02-20.htm

"Some of our islands..."
www.climate.org/topics/sealevel/index.shtml

64–65 CITIES AT RISK
Coastal populations
Gridded Population of the World (GPW), version 2 alpha: Consortium for International Earth Science Information Network (CIESIN), 1995.

Population Division of the Department of Economic and Social Affairs of the United Nations Secretariat, World Urbanization Prospects: the 2003 Revision, 2004.

New York
Athanasios Vafeidis, Flood Hazard Research Centre, University of Middlesex, personal communication
Goddard Institute for Space Studies, New York, NY: icp.giss.nasa.gov/research/ppa/2002/impacts/introduction.html

London
CEH, Edinburgh: www.nbu.ac.uk/iccuk/indicators/10.htm

Japanese ports
Mimura et al, quoted on Climate Change 2001: Working Group II: Impacts, Adaptation and Vulnerability, 1998: www.grida.no/climate/ipcc_tar/wg2/298.htm

Mumbai
Athanasios Vafeidis, Flood Hazard Research Centre, University of Middlesex, personal communication

Lagos
A C Ibe, L F Awosika. Sea level rise impact on African coastal zones in S H Omide and C Juma, eds, *A Change In The Weather: African Perspectives On Climate Change*, 1991: 105–12, African Centre for Technology Studies, Nairobi, Kenya: ciesin.columbia.edu/docs/004–153/004–153.html

66–67 CULTURAL LOSSES
Arctic
Arctic Climate Impact Assessment (ACIA), *Impacts of a Warming Arctic*, 2004, Cambridge University Press.

Boston, USA
B Fitzgerald, AD 2100: from Beantown to Water World? Geography profs: storm flooding to devastate Boston as globe warms, 25 March 2005, *BU Bridge*, VIII (24): www.bu.edu/bridge, accessed 23 June 2005.

P H Kirshen (Tufts University), W P Anderson (Boston University), M R (University of Maryland), Climate's Long-term Impacts on Metro Boston (CLIMB), media summary. Final report available from: www.tufts.edu/tie/climb

Belize Barrier Reef
Legal steps taken to protect natural heritage sites from climate change, 25 November 2004. www.iema.net, accessed 27 May 2005.

Scotland, UK
D J Breeze, Foreword, in Tom Dawson, ed, Conference proceedings: coastal archaeology and erosion in Scotland, *Historic Scotland*, 2003, Edinburgh.

Czech Republic
R Johnston et al, Czech Republic's cultural losses mount in flood's wake, 2002: portal.unesco.org

Mt Everest
Legal steps taken to protect natural heritage sites from climate change, 25 November 2004: www.iema.net, accessed 27 May 2005.

Alexandria, Egypt
M H Mostafa, N Grimal, D Nakashima, eds, Underwater archaeology and coastal management: focus on Alexandria, 2000, UNESCO, Paris.

M El-Raey, Impact of Climate Change on Egypt cesimo.ing.ula.ve/GAIA/CASES/EGY/impact.htm#Abstract

Thailand
M Draper, Floods threaten Thai monuments, *Archaeology*, 1996, 49 (2): www.archaeology.org/9603/newsbriefs/thai.html

West Coast National Park, South Africa
D Roberts, article in *The Cape Odyssey*, June/July 2002: www.sawestcoast.com/fossileve.html, accessed 23 May 2005.

Venice, Italy
Plumbing the depths, *New Scientist*, 2004, 183(2457): 36 (4).

Tuvalu
A Kirby, Pacific islanders flee rising seas, 9 October 2001: news.bbc.co.uk/1/hi/sci/tech/1581457.stm, accessed 23 May 2005.

Part 5: Responding to Change
Quote: http://news.bbc.co.uk/2/hi/science/nature/4508928.stm

70–71 INTERNATIONAL ACTION
www.unfccc.net
International Petroleum Industry, Environmental Conservation Association (IPIECA), Climate change: a glossary of terms, January 2001, 3rd edn: www.ipieca.org/downloads/climate_change/Glossary_3rd_edition.pdf
www.cei.org; www.api.org; www.iisd.org; www.neweconomics.org; www.climatenetwork.org; www.generationfoundation.org; www.greepeace.org; www.sidsnet.org/aosis; www.opec.org

72–73 MEETING KYOTO TARGETS
www.unfccc.int
European Commission, DG Environment, European Environment Agency, Annual European Community greenhouse gas inventory 1990–2003 and inventory report 2005, revised final version 1.3 submitted to the UNFCCC Secretariat 27 May 2005.

Counseil de l'union Europeenne, Communiqué de Press, 2647ème session du conseil, Environnement. Bruxelles, 10 March 2005, 6693/05 (presse 40).

Greenhouse gas emissions
Progress towards Kyoto targets
UNFCCC, Key GHG Data, 2005: unfccc.int/essential_background/background_publications_htmlpdf/items/3604.php

Long-term effect of CO_2 emissions
IPCC, Climate change 2001 summary for policymakers: www.ipcc.ch/present/graphics/2001syr/large/01.17.jpg

74–75 CARBON TRADING
Emissions reductions purchased
Emissions reductions supplied
Types of projects
Allowance-based exchange
F Lecocq, K Capoor. State and trends of the carbon market 2005.
A joint product of the carbon finance product of the World Bank
and of the International Emissions Trading Association: http://
carbonfinance.org/Router.cfm?Page=DocLib&ht=25621&dtype=25
622&dl=0 , accessed May 2006.

Project-based exchange
F Lecocq, K Capoor. *State and Trends of the Carbon Market*
2005, *State and Trends of the Carbon Market* 2003
F Lecocq, *State and Trends of the Carbon Market* 2004: http://
carbonfinance.org/Router.cfm?Page=DocLib&ht=25621&dtype=25
622&dl=0 , accessed May 2006

Project-based market
The World Bank Carbon Finance Unit Project Portfolio:
carbonfinance.org/Router.cfm?Page=ProjPort&ItemID=24702

76–77 FINANCING THE RESPONSE
Donors and recipients
World Resources Institute, Washington DC, Climate Analysis
Indicators Tool (CAIT) version 2.0: cait.wri.org/, accessed end
2005.
GEF project database: www.gefonline.org/home.cfm
Official Development Assistance increases further but 2006
targets still a challenge, OECD Development Assistance
Committee statistical sources: www.oecd.org
International Development Statistics: www.oecd.org

GEF climate change projects
Eberhard, A.A. and S.E. Tokle, with A. Viggh, A. del Manaco,
H.Winkler and S. Danyo, 2004. Climate change program study.
Washington, DC: Global Environment Facility
GEF project database: http://www.gefonline.org/home.cfm

78–79 LOCAL COMMITMENT
"We know the science..."
Arnold Schwarzenegger's weekly radio address, 11 June 2005:
www.schwarzenegger.com/en/news/uptotheminute/news_upto_
RadioAddress16.asp?sec=news&subsec=uptotheminute

Cities committed to change
Cities for Climate Protection: www.iclei.org/index.php?id=809,
accessed 23 November 2005.

Hyderabad, India
ICLEI, C-I, Traffic management measures for better environment:
a case of Hydrabad City, 2004: www.iclei.org/southasia/
CCP_Web_April2004/NewsLetter_june2004.PDF, accessed 22
September 2004

Finland
ICLEI in Europe, 26 January 2004, Finland: www.iclei-europe.
org/index.php?id=601move, accessed 22 Sept 2004.

Rayong, Thailand
Curitiba, Brazil
Cities for Climate Protection. http://www.iclei.org/co2/

Mareeba Shire County, Queensland, Australia
Cities for Climate Protection: local government action on climate
change, *CCP Australia Measures Evaluation Report 2005* http://
ccp.iclei.org/ccp-au/publication/290.pdf

USA
www.ci.seattle.wa.us/mayor/climate/

Regional Greenhouse Gas Initiative (RGGI)
www.rggi.org

80–81 CARBON DIOXIDE AND ECONOMIC GROWTH
H G Huntington. US carbon emissions, technological progress
and economic growth since 1870, Nature, 435, 2004, Energy
Modeling Forum: www.nature.com/nature/journal/v435/n7046/
full/4351152a.html

"There is an urgent need..."
Claude Mandil, International Energy Agency (IEA) at the launch
in Brussels of oil crises and climate challenges: 30 years of
energy use in IEA countries:
www.iea.org/Textbase/press/pressdetail.asp?PRESS_REL_
ID=1233/2/2004

Trends in carbon intensity
Carbon intensity
Change in carbon intensity
International Energy Agency, IEA Energy Information Centre,
on-line data services, http://data.iea.org/ieastore/default.asp

82–83 RENEWABLE ENERGY
European Wind Energy Association, Statistics: www.ewea.org/
index.php?id=180&no_cache=1&tsword_list[]=growth

Commitment to renewables
E Martinot, Global revolution: a status report on renewable
energy worldwide, *Renewable Energy World*, Nov–Dec 2005,
8 (6): 28, James & James (Science Publishers) Ltd, London.
Available as Renewables 2005 Global Status Report: www.
martinot.info/
OECD/IEA. Global Renewable Energy Policies and Measures
Database, 2006: http://www.iea.org/textbase/pamsdb/grindex.
aspx

Solar energy
Paul Maycock, PV NEWS annual review of the PV market 2004
as cited in European Photovoltaic Industry Association, Statistics
– world: www.epia.org/03DataFigures/DataWorld.htm, accessed
14 September 2005.
2004 figure from Paul Maycock, PV market update, *Renewable
Energy World, Review Issue 2005–06*, July–August 2005, 8 (4):
86, James & James (Science Publishers) Ltd, London.

Renewable energy sources
E Martinot, 2005 op cit

Renewable energy production
International Energy Agency, 2006 Gross Electricity and
Heat Generation from Renewable Sources (GWh, TJ) in the
Renewables Information (2005 edition) Accessed March 2006.
Renewable energy is a diverse and fast growing area. The data
reported here were gathered by an international agency using
a standardized questionnaire, but due to reporting issues, they
are not complete. We encourage people interested in specific
countries or technologies to consult additional sources, such
as Eric Martinot's *Renewables 2005 Global Status Report*, cited
above.

84–85 ADAPTING TO CHANGE
UNDP Disaster Risk Index: www.undp.org/bcpr/disred/english/
wedo/rrt/dri.htm
GE Source Natural Hazards: www.gesource.ac.uk/hazards/
A J McMichael et al, eds, Climate change and human health,
risks and responses, 2003, WHO, Geneva.
W N Adger et al, New indicators of vulnerability and adaptive
capacity, 2004, Tyndall Centre for Climate Change Research.
EM-DAT: The OFDA/CRED International Database:
www.em-dat.net.
N Brooks, W N Adger, Assessing and enhancing adaptive
capacity, 2005, technical paper 7, in B Lim, E Spanger-Siegfried,

eds, *Adaptation Policy Frameworks for Climate Change: Developing Strategies, Policies and Measures*: 266, Cambridge University Press.

R Few et al, Floods, health and climate change: a strategic review, 2004 (working paper 63, Tyndall Centre for Climate Change Research.

N Brooks, W N Adger, P M Kelly, The determinants of vulnerability and adaptive capacity at the national level and the implications for adaptation, *Global Environmental Change*, 15: 151–163.

Capacity to adapt
UNDP, New York, Human Development Report 2005: hdr.undp.org/statistics/highlights.

United Nations Development Programme, Bureau for Crisis Prevention and Recovery, *Reducing Disaster Risk, A Challenge for Development*. 2004, UNDP, New York. Data sourced from: EM-DAT: The OFDA/CRED International Database: www.em-dat. net/. See following for guidelines on data collection and recent changes in data recording: www.em-dat.net/guidelin.htm

Bangladesh
M Alam, Bangladesh country case study for National Adaptation Programme of Action (NAPA), workshop, 9–11 September 2003, Bhutan: www.unitar.org/ccp/bhutan/11–%20Bangladesh%20Case%20Study.pdf, accessed 22 November 2003.

oxfam.org.uk/coolplanet/teachers/disaster/wrksht3.htm
news.bbc.co.uk/2/hi/south_asia/503139.stm

Africa
www.fews.net

Part 6: Committing to Solutions
Quote: www.climatestar.org/stars.html

88–89 PERSONAL ACTION
coinet.org.uk
www.chooseclimate.org/flying/mapcalc.html
www.carbonfund.org/
Climate Care: www.co2.org/index.cfm
www.earthfuture.com/climate/calculators/
www.climatesolutions.org

Becoming carbon neutral
New Scientist 10 September 2005: 37–41.

90–91 PUBLIC ACTION
Corporate commitment brings increased profits
www.theclimategroup.org
www.theclimategroup.org/index.php?pid=592
Press release 2005: www.press.bayer.com/BayNews/BayNews.nsf/id/C125709F007F6298C125704A0029F331?Open&tccm=015030

Part 7: Tables
Quote: http://business.guardian.co.uk/story/0,3604,532566,00.html
Column 1, Population: World Bank Quick Reference Tables www.worldbank.org; CIA Factbook www.cia.gov
Column 2, GNI per capita: World Bank Quick Reference Tables www.worldbank.org; CIA Factbook www.cia.gov
Column 3, Human Development Index: UNDP, New York, Human Development Report 2005: hdr.undp.org/statistics/highlights.
Column 4, Water withdrawn: World Bank: *World Development Indicators 2005*, supplemented with data from Table: Freshwater Resources 2005, EarthTrends, World Resources Institute, earthtrends.wri.org/
Column 5, Coastal population: Gridded Population of the World (GPW), version 2 alpha: Consortium for International Earth Science Information Network (CIESIN), 1995.
Column 6, Weather-related disasters: EM–DAT: The OFDA/CRED International Database: www.em-dat.net.
Column 7, Carbon dioxide emissions: Emissions from burning fossil fuels (annual total and tonnes per person): International Energy Authority: www.eia.doe.gov/iea/carbon.html
Million tonnes 1950–2000: World Resources Institute, Washington DC, Climate Analysis Indicators Tool (CAIT) version 2.0: cait.wri.org, accessed 20 June 2005.
Transportation: International Energy Agency, 2006, CO_2 Emissions from Fuel Combustion (edition 2005), Paris, France, http://data.iea.org/stats/eng/main.html
Column 8, Methane emissions: World Resources Institute, CAIT version 3.0, 2006, Washington DC. World Resources Institute, Washington DC, Climate Analysis Indicators Tool (CAIT) cait.wri.org
Column 9, Carbon intensity: World Resources Institute, CAIT version 3.0, 2006, Washington DC. World Resources Institute, Washington DC, Climate Analysis Indicators Tool (CAIT) cait.wri.org

Photo credits
The publishers are grateful to the following for permission to reproduce their photographs:
18 Larsen B ice shelf, NASA; 22 Larsen B ice shelf, European Space Agency; 23 Arctic ice cap: NASA; Greenland ice cap: courtesy of Greenland Melt Extent, 2005, Steffen Research Group, Cooperative Institute for Research in Environmental Sciences (CIRES), University of Colorado at Boulder http://cires.colorado.edu/science/groups/steffen/greenland/melt2005; 24 Arapaho glacier: 1898 R.S. Brackett, in R S Waldrop, *Arapaho Glacier: A Sixty Year Record*. University of Colorado 1964 and 2003 Tad Pfeffer; 25 Glacial lakes, Bhutan: Jeffrey Kargel, USGS/NASA JPL/AGU; 27 European heatwave: Reto Stöckli, Robert Simmon and David Herring, NASA Earth Observatory, based on data from the MODIS land team; 28 Jason Verschoor/iStockphoto; 34 Ice core: http://cfa-www.harvard.edu/press/pr0310image.html; 37 Globes: David Stainforth et al at www.climateprediction.net; 38 Steve Lovegrove/iStockphoto; 42 Coal: Rulmeca Corp; 44 Methane hydrates: courtesy of Ian R. MacDonald, Texas A&M Univ. Corpus Christi, http://www.onr.navy.mil/media; 51 Supermarket USDA/David F. Warren; Brussel sprouts: Amanda Rohde/iStockphoto; 52 Paulus Rusyanto/iStockphoto; 54 Tiger: Stefan Ekernas/iStockphoto; Beach: Marcus Tuerner/iStockphoto; Golden Pagoda: Craig Hilton Taylor/IUCN Red List Programme; 60 Flood: WHO; 62 Tuvalu: Mark Lynas; 66 Caribou: Carol Gering/iStockphoto/; Boston: Joshua Velasquez/ iStockphoto; Belize Barrier: Dennis Sabo/iStockphoto; 67 Czech Republic: Sándor Szabó/ iStockphoto/; Mt. Everest Peter Hazlett/ iStockphoto/; Thailand: Ine Beerten/iStockphoto/; Venice: Chris Klein/ iStockphoto/; Tu valu: Mark Lynas; 68 iStockphoto; 75 top: World Bank Carbon Finance Unit; bottom: Nordex; 83 Small hydro: World Bank/Dominic Sansoni; Wind: Jason Stitt/ iStockphoto; Biomass: NREL/ Warren Gretz; Geothermal: NREL/Joel Renner, INEEL; Solar: NREL/ Robb Williamson; Tide, wave, ocean: Sustainable Energy Forum 84 left: Rebecca Gustafson/ USAID; 86 Amici della Terra; 90 Global Images; 92 Marek Pawluczuk/iStockphoto

Index

action 88–89, 90–91
adaptation 10, 11, 12, 17, 70, 76–77,
    84–85, 91
    Adaptation Fund (AF) 76
Africa
    crop yields 10, 59
    deforestation 48
    drought 21
    farming practices 48
    floods 21
    glaciers 25, 55
    Lagos 65
    precipitation 10, 33, 37
agriculture 41, 48, 50–51, 53, 54,
    58–59, 62
    as percentage of GDP 50–51
    emissions from 44, 50–51
albedo 30
Alliance of Small Island States
    (AOSIS) 70–71
allowance-based markets 74, 75
alternative energy see energy,
    renewable
animal behavior 9, 20, 22, 66
Annex I countries 16, 72, 73
Antarctic 20, 22–23, 36
    ice sheets 22
    Peninsula 20, 24
Arctic 20, 22–23, 36
    ice cap 23, 53
    indigenous peoples 23, 66, 84
    shipping route 23, 53
Asia
    deforestation 48
    floods 21
    glaciers 25
    heatwave 21
    precipitation 43
    rice production 50
Asia-Pacific Partnership 76
atmospheric
    circulation 32
    composition 9, 20, 29, 30, 34, 36,
        58
    lifetime 31
Australia
    carbon trading 74
    drought 21, 33
    heat 26
    Kyoto Protocol 69, 70, 72, 73
    local mitigation plans 12, 69, 79
    project-based exchange 74
    renewable energy 83
aviation 41, 46, 47

Bangladesh
    cyclone shelters 85
    Sundarbans Delta 55

biodiversity 54
biomass fuel 75, 83
Bolivia 20
Brazil
    carbon dioxide emissions 43
    Curitiba 78
    droughts 20
    floods 20
    GEF funding 77
    hurricanes 20, 26
    project-based exchange 74
    savannah 54

Canada
    polar bears 20
    project-based exchange 74
    Rockies 24
carbon
    credits 74
    exchange 48, 49
    footprint 88
    intensity 80–81
    neutral 89
    sinks 14, 48, 54
    trading 12, 69, 74–75
carbon dioxide
    and economic growth 80–81
    atmospheric concentration 9, 30,
        31, 34, 58, 72
    atmospheric lifetime 31, 40
    cumulative emissions 40–41
    from agriculture 50
    from fossil fuel burning 40–41,
        42–43
    global warming potential 44
    in ice-cores 34
    in oceans 33
    long-term effect 72
    past levels 34, 35
    per person emissions 43
    personal emissions 88–89
carbon tetrachloride 31
Caribbean
    hurricanes 20, 26, 33
    islands 62
caribou 66
Central America, endemic mammals
    54
Central Asia, mountains of 25, 55
cereal production 58–59
Chicago Climate Exchange (CCX)
    75
China
    carbon dioxide emissions 43
    carbon intensity 39, 80
    coal 42, 43
    drought 21
    floods 21

GEF funding 77
    solar thermal energy 82
Cities for Climate Protection (CCP)
    78–79
cities
    at risk 62, 64–65
    taking action 12, 78–79
Clean Development Mechanism (CDM)
    14, 74, 76–77, 80
Climate Action Network 70
climate change
    convention see United Nations
    economic costs 66
    future 34, 36–37
    history 24, 34–35
    projections 17, 36–37
    signs 20–21, 22–23, 24–25
climate system 32–33, 36
coal 39, 42, 48
    mining and methane production
        44
    coastal erosion 64, 65, 66
    storms 64, 65, 66
Colombia 75
Competitive Enterprise Institute 70
computer models 9, 29, 34, 35, 36
Convention on Climate Change 70–71,
    76, 78
coral bleaching 15, 21, 66
corporate commitment 12, 90
cultural losses 66–67
cyclones 36, 84
Czech Republic, floods 67

deaths
    from disease 60–61
    from respiratory conditions 60
    from weather-related disasters 21,
        26, 84–85
deforestation 30, 39, 54
disease 20, 60–61, 64
    in plants 58
drought 26, 32, 54, 58, 60, 84
    Africa, Horn of 21, 84
    Australia 21
    Spain 21

economic development 10, 36, 39, 53,
    65, 70, 80–81
economies in transition (EITs) 70, 72,
    73, 74
ecosystems 15, 10, 20–21, 48, 54–55,
    70
Egypt 62, 67
El Niño 33
energy
    efficiency 12, 75, 76, 80, 87, 88
    production 39, 40, 41, 42, 44, 64

renewable 11–12, 39, 42, 69, 75, 82–83, 87
Environmental Integrity Group (EIG) 70–71
Europe
  Alps 21, 25
  birds 55
  butterflies 21
  floods 26, 67
  heatwave 10, 21, 27, 33, 74, 87
  land-use change 48, 49
  plants 21
European Union (EU) 16, 71, 72
  Emissions Trading Scheme (ETS) 12, 74, 75
extinction of species 54–55
extreme weather see weather, extreme

Famine Early Warning System 84
farming see agriculture
feedbacks 30
Finland 79
fires 21, 60
fisheries 33
floods 26, 36, 56, 84
  Argentina 20
  Brazil 20
  effect on health 60–61
  Ethiopia 21
  Europe 67
  New Zealand 21
  South Asia 21
food
  production 60, 70
  security 11, 58
  shortage 56
  transportation of 51
forests 54, 75
fossil fuels 12, 48
  burning of 30, 36, 40–41, 42–43
  efficient use of 42, 75, 76, 80

G8 countries 16, 41, 46
gas 48
Generation Foundation 70
geothermal energy 15, 82, 83
Germany 83
glaciers 15, 24–25
  calving 22
  melting 54, 62
  retreating 20
  thinning 20
Global Environment Facility (GEF) 76–77
global warming 35, 33, 37, 40
  effect on ecosystems 54
  potential (GWP) 15, 44–45
Global Wind Energy Council 82

greenhouse effect 29, 30–31, 34
  greenhouse gases 15, 31, 36
  atmospheric concentration of 20, 29, 30, 34, 36, 39, 58
  capture of 75
  from agriculture 39, 50–51
  from energy production 39
  from fossil fuel burning 40–41, 42–43
  from transportation 39, 46–47
  future emissions 40
  growth in emissions 41, 46, 47
  human-induced 9, 29, 31, 34, 35, 70
  natural 34, 35, 48
  past emissions 12, 40–41
  reducing emissions 11, 69, 72–83, 88–91
  scenarios of future emissions 17
  sectoral breakdown 41
Greenland ice cap 22, 23, 24, 62
Greenpeace 70
gulf stream 33

halocarbons 14, 31, 44
health 60–61
heatwaves 21, 26, 27, 32, 33, 84
Himalaya 25, 67
Honduras 75
Human Development Index 85–85
humidity 36, 60
hurricanes 20, 26, 33, 64, 66
hydroelectric, small-scale 75, 76, 83
hydrofluorocarbons 44, 45, 75

India
  carbon dioxide emissions 43
  carbon intensity 39, 80
  farming practices 48
  GEF funding 77
  heatwave 21
  Hyderabad 79
  Mumbai 26, 65
  project-based exchange 74, 75
  Sundarbans Delta 55
Indonesia
  drought 33
  GEF funding 77
  glaciers 25
industrial processes 30, 41, 42, 44, 64
Industrial Revolution 31, 39
industrialization 30, 36, 40, 41, 80
industrialized countries 10, 12, 40–41, 42, 70
inequality 39, 84–85
Intergovernmental Panel on Climate Change (IPCC) 13, 15, 29, 34, 54
International Council for Local

Environmental Initiatives (ICLEI) 78
International Energy Agency (IEA) 82
International Institute for Sustainable Development (IISD) 70
intestinal diseases 60–61
Inuit Circumpolar Council 23
irrigation 56, 64
Italy
  Rome 79
  Venice 67

Japan 65
  carbon dioxide emissions 43
  ports 64
  project-based exchange 74
  renewable energy 83
jet stream 32
Joint Implementation (JI) 15, 74
Joint Science Academies 20

Kyoto Protocol 16, 46, 69, 70–71, 72–73, 74, 87

land-fill sites and methane production 44
  gas capture 75
land-ice, melting 62, 63, 72
land-use change 10, 41, 49
Larsen ice shelves 20, 22
Least Developed Countries (LDCs) 16, 41, 70
  Fund 76
livestock farming 30, 44, 50

malaria 60–61
Maldive islands 62
malnutrition 58–59, 60–61
manufacturing and construction 41
Mauna Loa observatory 19
methane 30, 35, 44–45
  atmospheric lifetime of 31
  from agriculture 44, 50
  from fossil fuel burning 42
  hydrates 44
Mexico
  GEF funding 77
  glacier 24
monitoring systems 19
mosquitoes 20, 60, 61
motor vehicles 46, 47

National Adaptation Plans of Action (NAPA) 76–77
natural gas 42
New Economics Foundation 70
New Zealand
  floods 21

glaciers 25
    project-based exchange 74
newly industrializing countries 42
nitrogen-based fertilizers 44, 50
nitrous oxide 31, 42, 44, 50, 75
North America
    land-use change 48, 49
    mosquitoes 20
North Pole 22, 23
nuclear power 11

ocean
    carbon exchange 48
    circulation 33
    energy source 82, 83
    salt concentration 33
    temperature 33
    thermal expansion 62, 63, 72
    warming 22, 55
oil 42, 48
Organisation for Economic
    Co-operation and Development
    (OECD) 16, 74, 82
Organization of Petroleum Exporting
    Countries (OPEC) 16, 70–71
ozone 60

Pacific islands 54, 62, 63, 67, 87
paleoclimatology 34–35
perfluorocarbons 14, 44, 45
permafrost melt 30
    Alaska 20
    Siberia 21
pests 58
photovoltaics 16, 75, 76, 82, 83
plants 20
    Antarctic 22
    Arctic 66
    Europe 21
    migration of 9, 54, 66
    role in carbon cycle 48
    USA 20
polar bears 20, 66
population growth 36, 65
ports 53, 65
poverty 60
    eradication 12, 57
    people living in 10, 53, 64
precipitation 16, 20, 21, 24, 36
    change 10, 37
    heavy 20, 21, 26
    patterns 32, 58, 60
project-based markets 74, 75

radiative forcing 16, 30, 31
rainfall see precipitation
renewable energy see energy,
    renewable

river flows 54, 55
Russia
    carbon dioxide emissions 43
    Siberian permafrost melt 21

saltwater contamination 64, 65
sanitation systems 64
Scandinavia 25
sea-level rise 22, 53, 54, 55, 62–63,
    66, 72
    and cities 64–65
Senegal, carbon dioxide emissions 43
shipping 41, 47
soil 48, 58
soil-borne diseases 61
solar energy 30, 32, 34
solar thermal energy 82, 83
South Africa 55, 67
South America, deforestation 48
Spain
    drought 21
    Imperial Eagle 55
    renewable energy 83
Special Climate Change Fund (SCCF)
    76
storm surges 62, 63
Strategic Priority: Piloting an
    Operational Approach to
    Adaptation (SPA) 76
sulfur hexafluoride 14, 44

temperature
    change 54
    increase 20, 22, 23, 30, 37, 45,
    54, 58
    land surface 27
    ocean 22, 26
    patterns 32, 34, 35, 37, 60
Thailand 67, 79
thermohaline circulation 16, 33
tick-borne diseases 60
trade 10, 46, 47, 51, 53
transportation 10, 41, 42, 46–47, 69,
    88–89
tropospheric ozone 14, 30, 31

UK
    carbon dioxide emissions 43
    London, threat to 65
    project-based exchange 74
    renewable energy 83
    Scotland 55, 67
Umbrella Group 70–71
United Nations Framework
    Convention on Climate Change
    (UNFCCC) 12, 16, 70–71, 76, 78

urban
    development 12, 46, 54
    dwellers 11, 64
USA
    agriculture 53
    Alaska 9, 20, 23, 24, 87
    Boston 66
    carbon dioxide emissions 43
    carbon intensity 80
    carbon trading 74
    emissions reduction 12, 69
    glaciers 24
    hurricanes 20
    Kyoto Protocol 11, 70, 72, 73
    Mayors' Climate Protection
      Agreement 78
    mosquitoes 20
    New Orleans 11, 64, 84
    New York 64
    plant behavior 20
    precipitation 26
    project-based exchange 74
    red fire ant 54
    Regional Greenhouse Gas Initiative
      12, 78
    renewable energy 83
    transportation 46
    UNFCCC 70
    water supplies 62
utility companies 82

volatile organic compounds 60
volcanic eruptions 34
vulnerability 17, 54–67, 84–85

waste 41, 89
water
    contamination of 64, 65
    stress 24, 56–57
    supplies 62, 64
    use 56
    withdrawals 57
waterborne diseases 60
water-intensive industries 56
water-saving technologies 54
water vapor 15, 30
weather
    extreme 20–21, 26–27, 36
    unpredictable 20, 56
weather-related disasters 26–27, 77,
    84–85
Web of Science 34
wind power 75, 76, 82, 83
windstorms 20, 26, 33, 36, 62, 64, 84
World Bank, Community Development
    Carbon Fund 75
World Health Organization 60